The Packaging Designer's Book of Patterns

The Packaging Designer's Book of Patterns

Second Edition

László Roth and George L. Wybenga

JOHN WILEY & SONS, INC.

Library of Congress Cataloging-in-Publication Data:

Roth Lászlo.
   The packaging designer's book of patterns / Lászlo Roth and George L. Wybenga.—
Second edition
      p. cm.
   Includes bibliographical references and index.
   ISBN 0-471-38504-2 (pbk.)
   1. Packaging—Design.   2. Paperboard.   I. Wybenga, George L.   II. Title.

TS195.4.R683 2000
688.8—dc21

                                                                    00-022822

Printed in the United States of America.

10 9 8 7 5

Contents

Preface

Ten years ago, when we started this project, the prospect of delivering over 400 packaging patterns to the publisher seemed overwhelming. However, the manuscript was delivered on time, and in the fall of 1991 we excitedly perused the results of our labor. Initially conceived as an aid to our design students at the Fashion Institute of Technology in New York, we did not expect the world-wide acclaim this manual engendered.

Over the years the publisher approached me several times with a request for a revised edition. I was reluctant to undertake this task mainly because I thought that it would be impossible to add a sufficient number of new patterns to make it worthwhile for both designers and students. Then several errors were pointed out to me by the users of this book. The necessary corrections finally convinced me that it was time to accede to the publisher's request. I set out to collect and develop the 45 new patterns called for by the publisher. 100+ patterns later I decided to stop in order to meet the publication deadline. In addition to the corrections and the new patterns, all the glue areas have been indicated with black dots.

Some reviewers have suggested that environmental concerns be stressed. In the preface of the original volume we addressed that subject briefly. Packaging is a must for getting the product safely to the consumer. The purchaser of a product has the right to be assured that the merchandise is whole, unpolluted, and untampered with. In western European countries the producer is responsible for the cost of recycling any materials other than paperboard incorporated in the packaging (e.g., aluminum or plastic pour-spouts, styrene platforms, etc.). All constructions in this book are made exclusively of paper and are completely recyclable. Many packaging configurations lend themselves to after-use as storage bins, keepsake and collectors' items, or play-value.

The various sections of this book only serve an organizational purpose. Thus, a paperboard carton may be easily converted into a corrugated container, and vice versa. The many different tabs, locks, closures, etc. are often interchangeable and may be adapted to the container's requirements and/or the designer's preference. As stated before, some configurations or machinery may be protected by one or more patents. Each pattern in this volume has been photocopied on oaktag and constructed in order to verify the dimensional accuracy and as a model for the three-dimensional drawings of the containers. All patterns are proportionally correct but may be altered to the need of the product to be packaged. The thickness of the paper- or corrugated board will be determined by the weight, fragility, etc. of the product to be packaged.

I would like to thank Susanne Anoushian, Martin R. Carbone, Elizabeth Downey, and Marianne Rosner-Klimchuk for their invaluable suggestions; John Edminster for his eagle eye and contribution to many new Point-of-Purchase display patterns; Max Hilaire for resupplying me with the authors' photograph; Amos Glen, manager of the King Kullen supermarket, for his courtesy; and Margaret Cummins, Jim Harper, and Maury Botton of John Wiley & Sons for guiding me through the process of this project.

I had fun playing with paperboard and manipulating scores and cuts to see what different forms I could develop. It is hoped that his new edition will continue to be enthusiastically received and be useful in its purpose. As you explore the many exciting shapes packaging can take, think of the ingenuity of the structural designers who initially came up with some of the configurations and their functions.

George L. Wybenga

Preface from the First Edition

Here is the first definitive collection of patterns and structural designs for paperboard packaging, point-of-purchase displays, and other three-dimensional graphic products. This workbook contains over 450 patterns and structural designs that can be adapted for packaging. The designer may choose among hundreds of alternative carton and box structures, which may be used as they are, modified, or adapted to create exciting new structures. The patterns are accompanied by a description of the historical and technical origins of the materials and methods used. Special attention is given to computer-aided design as an engineering tool that offers new insights into carton and box design and manufacturing.

Paperboard packaging has a profound impact on the U.S. economy. In 1988 about $16 billion was spent for paperboard packaging, accounting for 45 percent of the total expenditure for packaging. Today the United States leads the world in the use of paper and paperboard, with a consumption of over 600 pounds per person per year. There are more than 5000 plants in the United States that manufacture and convert paper and paperboard. In addition, there are about 530 carton manufacturers with 752 plants and a total employment of 80,000. The corrugated board industry includes 795 companies with 1427 plants and a total employment of 118,000.

The packaging materials industry serves every major retail environment—in many ways it has shaped the American lifestyle. If it is to continue to thrive, it must be creative and innovative. It must be market oriented and must continually supply new methodologies and new solutions to environmental problems.

Along with sensitivity to environmental problems, creativity should be the designer's most important consideration. This workbook may be used not only as a reference for structural design but also as a source of exercises in learning to stimulate creative skills. The cartons shown in this book are designed to depict generic styles whenever possible. Each illustration is accompanied by the name that is commonly used to describe the style. The grain direction of all folding cartons is horizontal unless otherwise indicated. The flute direction of all corrugated cartons is vertical unless otherwise indicated.

Some of the packages currently in use are patented or have patent-protected features for use on proprietary packaging machines. Where appropriate, patented features are indicated. The packages that appear on the following pages are patented or contain patented features: 226, 237, 269, 278, and 360.

The reader is advised that patented features are protected by patent law and may not be used in practice without the express permission of the patent holder. For information about patents, contact the U.S. Patent Office.

ACKNOWLEDGMENTS

This book would not have been possible without the generous help, contributions, and advice of our colleagues in the packaging industries. We are especially indebted to Robin Ashton, editor-in-chief of *Packaging*, a Cahner's publication, for allowing us to adapt some of the diagrams and charts from the *Packaging Encyclopedia and Technical Dictionary*. Our special thanks also go to the following packaging professionals and associations:

National Paperbox & Packaging Association

Bruce Benson, Fibre Box Association

S. Edward Iciek, Paperboard Packaging Council

R. Troll, Association of Independent Corrugated
    Converters

D. F. Zulke, The Langston Corporation

James S. Solch, Book Covers, Inc.

Professor John Barrett

Professor Jack Schecterson

Professor Anita Mott

Professor Marianne Rosner-Klimchuk

Our thanks also to photographers Max Hilaire and Charles Stewart.

Finally, we wish to thank Lilly Kaufman, Amanda Miller, and Liz Geller of Van Nostrand Reinhold for their patience and encouragement in our effort to make this book an important contribution to the field of packaging design. Our thanks to them and to all those at Van Nostrand Reinhold whose assistance made this book possible.

László Roth
George L. Wybenga

The Packaging Designer's Book of Patterns

Introduction

17th Century Dutch Paper Mill

Paper is among the noblest of human inventions. It is worthwhile, therefore, to begin with a short history of papermaking.

Before books could be written and preserved, a writing surface had to be developed that was light, not too bulky, and was easily stored. The first great advance was the Egyptians' use of papyrus in the third millennium B.C. Sheets of beaten papyrus stems were fastened together into scrolls, some more than 120 feet long, that could be rolled up for storage. After papyrus came parchment, which was perfected in Asia Minor in the city of Pergamum (from which its name is derived) in the second century B.C. Animal skins had long been used as a writing surface in Greece and Rome, but it was in Pergamum that methods were evolved for the production of a durable, velvet-smooth parchment that could be written on on both sides.

For hundreds of years all paper was made by hand from rag pulp. The use of wood fibers to make paper was discovered in the mid 1800s. In 1840 Friedrich G. Keller in Germany invented a way to grind logs into a fibrous pulp; this method produced a rather poor quality of paper, as all parts of the wood—not just the fibers—were used.

Paper as we know it today was first made in China in 105 A.D. Ts'ai Lun, a member of the court of Emperor Ho Ti, succeeded in turning husks of cotton fibers into paper pulp. This method spread throughout China, Korea, and Japan and as far west as Persia. In 751 A.D. Moslems captured a Chinese paper mill in Samarkand and learned the method of papermaking. They brought the method to Spain around 950 A.D., and by the thirteenth century paper mills had been established throughout western Europe, first in Italy and then in France, Germany, England, and Scandinavia.

The first paper mill in America was built in 1690 by William Rittenhouse near Philadelphia. Sheets of paper were produced one at a time until 1799, when Nicholas Louis Robert developed a continuous process. (This method was patented in England by the Fourdrinier brothers and is known by that

Detail of an Egyptian papyrus scroll (ca. 2500 B.C.).

Egyptian scribe using paper made from papyrus.

Early Chinese print on paper (ca. 300 A.D.).

name.) In 1817 the first cylinder-type papermaking machine, which can produce a better quality of paper in a continuous process, was invented by John Dickenson.

MODERN PAPERMAKING

Today almost all paper is manufactured from wood. Cellulose fibers (which account for 50 percent of the content of wood) are the primary ingredient, followed by lignin (about 30 percent), which acts as a fiber binder or glue.

Water plays an important role in modern papermaking. The manufacture of 1 ton of paper requires about 55,000 gallons of water, most of which is recycled. The papermaking process also uses sulfur, magnesium, hydroxide, lime, salt, alkali, starch, alum, clay, and plastics (for coating). There are two basic types of paper: fine paper for writing and paper for printing and industrial use (packaging).

The first step in manufacturing paper from wood is to remove the bark. The cheapest way to separate the fibers is to grind up the wood by forcing the logs against grindstones submerged in water. The water carries off the wood fibers. In this

Diagram of a papermill (greatly simplified). The pulp is ejected in a thin layer onto the conveyer-sieve (A). The pulp is then pressed through a great number of cylinders (B) and dried by a series of heated cylinders (C). It is then calendered (D) and taken up by the web (E).

process everything is used, and the paper produced is of low quality. Another, more frequently used process is chemical pulping, in which the wood is chipped into small pieces, the fibers are extracted through a chemical process, and the unusable material is eliminated. Chemical pulping is more expensive, but it produces better-quality paper.

Chemical pulping creates a pulp, which is then refined by washing and separating the fibers. Refinement, a time-controlled process during which the manufacturer can add various chemicals to increase bonding, texture, and water resistance, increases the quality and strength of the paper. Pigments (for coloring) and coatings (plastics) can also be added at this stage.

Once the pulp is prepared, it goes to one of two types of machines: The Fourdrinier, or the cylinder machine. Modern papermaking machines are huge. They can be as long as a city block and several stories high. They produce paper up to 30 feet wide at a speed of 3,000 feet per minute, resulting in 800 miles of paper a day! The primary papermaking machine is the Fourdrinier. Most Fourdrinier machines make only one layer of material, although they can be equipped to make several layers.

Paper produced by a Fourdrinier machine is smoothed by a stack of highly polished steel rolls, a process known as calendering. The finished paper is then cut, coated, and laminated.

Another frequently used papermaking machine is the cylinder machine. This machine makes heavy grades of paperboard, generally using recycled paper pulp. The pulp is built up in layers. Since paperboard is much thicker than paper, the drying operation is far more extensive. Large steam-heated cylinders drive the excess moisture out of the paper. A coating is then added to create a smooth surface.

The great advantage of the cylinder machine is that it uses large amounts of recycled paper in thick layers to provide strength.

Paper is bought on the basis of the weight (or basis weight), in pounds, of a ream of paper. (A ream is equal to 3,000 square feet of surface.) The thickness of paperboard is expressed in caliper points, which are stated in thousandths of an inch (usually written in decimals). Since most papers are laminated or coated with other materials, caliper points are rarely used today to specify weight. The paperboard used in folding cartons is specified according to the size of the carton or, more often, the weight of the item that goes into it. A glass bottle for 3.5 fluid ounces of fragrance, for example, would require a folding carton with a thickness of approximately 18–24 points.

The thickness of paper can be controlled by means of calendering, pressing, and laminating. High-quality paper is up to 12 points thick; paperboard varies in thickness from 12 to 70 points.

About 20 million tons of fine papers are used for printing and writing annually. Five and a half million tons are used for packaging. Tables 1, 2, and 3 list the major boxboards and papers used in packaging. The uses, content, and characteristics of these packaging materials are described.

PRINTING, FINISHING, AND DIE-CUTTING

Folding cartons are manufactured using three main processes: printing, die-cutting, and finishing.

Printing Methods

Several methods of printing are available, they include letterpress, offset lithography, gravure, flexography, and silk screen. Each method is suitable for particular types of jobs.

The *letterpress* method transfers ink from a metal plate directly to the sheet paperboard. This is one of the oldest methods of quality printing. New technologies have rendered it almost obsolete.

Letterpress: The ink is applied to the raised surfaces of the plate and transferred to the substrate.

Offset lithography has replaced letterpress because of its production efficiency and high-quality color reproduction. New high-speed presses and computer-aided systems, along with technological advances in inks and coatings, have made "offset" the most popular process for printing on folding cartons. In this process specially sensitized metal plates are chemically treated to accept ink. The ink is transferred from the plate to a smooth blanket roller, which then transfers the image to the paperboard.

    Gravure printing is used for high-quality reproduction in large-quantity runs (i.e., millions of copies). Specially etched printing cylinders have cells that accept and store inks. A "doctor blade" wipes off excess ink as the cylinder rotates to the impression cylinder, where the plate cylinder transfers the image to the paperboard. Gravure printing can be accomplished on an in-line web

Offset lithography: The ink is picked up by the pre-wetted plate, which transfers the inked image to the offset blanket, which in turn transfers the image onto the substrate.

Gravure (Intaglio): The ink is applied over the entire plate, scraped off the surface by the doctor blade, and transferred from the ink wells onto the substrate.

press, which is known as *rotogravure*, or on a sheet-fed press, which is called *photogravure*.

Flexographic printing is similar to letterpress printing. It uses a raised positive composition plate made of rubber or plastic. High-speed in-line web presses are used. This process has been associated with low-quality simple line art printing, but recent technological breakthroughs with fast-drying inks have made flexographic printing a low-cost, high-quality method for medium production runs.

Silk screening is a simple method of color printing in which a fabric mesh stretched over a frame is used instead of a printing plate. A stencil-type design is adhered to the mesh and pigment is "squeegeed" through the stencil. A separate "stencil" is required for each color used. (See diagram.)

Printing technology has changed in recent years to better meet the needs of carton manufacturing. Special coatings, varnishes, lacquers, and inks are available to give a bright finish or provide a moisture-proof barrier. Environmental problems have been alleviated by the introduction of water-based coatings and inks.

Die-cutting

The process of *die-cutting* involves creating shapes using cutting and stamping dies. There are three methods of die-cutting. *Hollow* die-cutting is done with a hollow die, which looks like a cookie cutter. This method is used exclusively for labels and envelopes. *Steel rule* die-cutting is used when a close register is required. Steel rules are bent to the desired shape and wedged into a ¾" piece of plywood. The die is locked up in a chase on a platen of the die-cutting press. Several sheets can be cut at once. A flatbed cylinder press can also be used for die-cutting.

The third method of die-cutting uses *lasers*, which were invented by C. H. Townes and Arthur Schawlow in 1958. (The word *laser* is an acronym for "light amplification by simulated emission of radiation.") The laser beam, which can be concentrated on a small point and used for processes such as drilling, cutting, and welding, has become widely used in manufacturing, communications, and medicine. Since a laser beam is extremely sharp and precise, the resulting cut is very accurate and clean.

Silkscreen: The ink is squeegeed through a stencil adhered to the mesh of the silk onto the substrate.

Laser cutout.

Steel rule die. This is a typical sample maker's die for a reverse-tuck carton. The die consists of scoring and cutting rules inserted into a sheet of plywood. A production die will have a great number of box layouts ganged so as to create as little waste as possible. Production dies are on either a flat sheet or a cylinder.

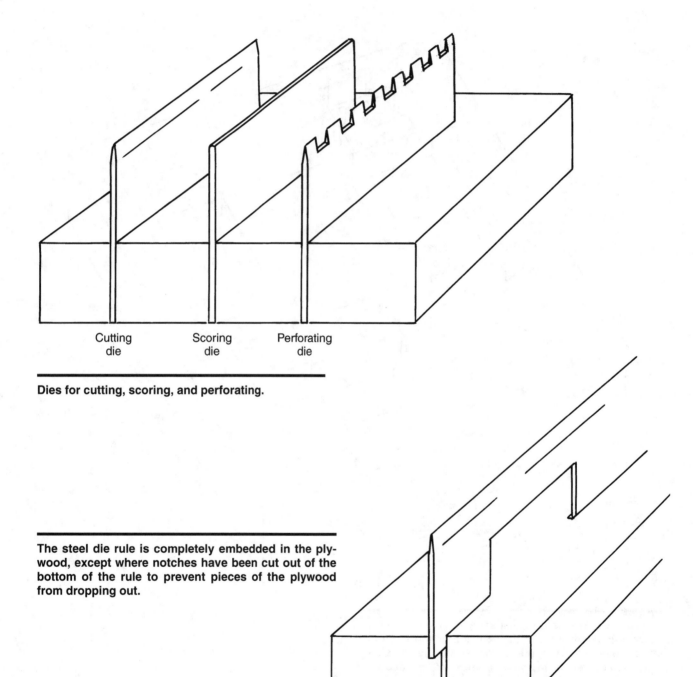

Cutting die   Scoring die   Perforating die

Dies for cutting, scoring, and perforating.

The steel die rule is completely embedded in the plywood, except where notches have been cut out of the bottom of the rule to prevent pieces of the plywood from dropping out.

Windows may be die-cut out of the face panel or extended into the top or side panels. They may be made in any shape desired for the final graphic solution. When laying out the window shape, consider leaving several shapes that may be printed with seasonal messages and retail price suggestions when the package blank is printed and can be removed when not required.

15°

1/2

1/2

Embossing. Paper and board lend themselves to *embossing*, the process by which a design image is made to appear in relief. Embossing can be superimposed on printing or done on blank paper (blind embossing) for a sculptured three-dimensional effect. It is achieved by pressing a sheet of paper between a brass female die and a male bed, or counter, both of which are mounted in register on a press. Embossing is generally used on prestigious packages; on packaging for cosmetics, gifts, and stationery; and on promotional materials.

Finishing

Finishing operations include gluing (using adhesives), windowing (die-cutting), coating, and laminating.

There are several types of gluing equipment, each designed for specific purposes. Right-angle gluers are used for Beers-style and six-corner construction; straight-line gluers are used for side seams and automatic bottom-closured cartons. Automatic tray-forming devices are used in the manufacture of tapered trays, clamshells, and scoop-style cartons. Today computer-controlled multipurpose gluers are used in most large paperboard-manufacturing plants.

A wide variety of specially formulated adhesives are used for specialty cartons made of coated, laminated, and plastic materials. Environmental conditions such as high moisture levels, freezing temperatures, sterilization, and microwaving require specific types of adhesives. Many types of adhesives are available, including self-adhesive and pressure-sensitive labels, resin emulsion adhesive for coated boards, and cold- or hot-melt adhesives for plastics.

Scoring. Paper and board come in flat sheets or rolls. In all papers and boards the fibers are aligned in one direction, called the *grain*. If they are torn with the grain, the edges will be smooth. If they are torn against the grain, the edges will be ragged.

When scoring the fold lines on the comprehensive, only *compress* the fibers of the paper, do not crush or cut them. Use a dull knife or letter opener. With practice you should be able to score on one side of the paperboard and achieve clean folds. Use white glue sparingly on the glue flaps. A thin coat of rubber cement may be used for covering papers, foils, or fabrics. *Caution:* rubber cement is highly flammable and toxic; library paste is the preferred alternative.

Scoring

Cutting. To facilitate the preparation of a packaging comprehensive, lay out the pattern carefully with triangle and T-square on the appropriate paper stock. Two-ply or three-ply are ideal for comps. Use a resilient backboard to score and cut; the chipboard on the back of drawing pads or a self-healing cutting board are great for this purpose. Use a steel straightedge for straight lines. To prevent sliding, glue a few strips of fine sandpaper to the back of the straightedge. An X-Acto® knife is easy to control for both straight and curved lines. Keep fingers well away from the cutting blade.

Cutting

11

Table 1. BOXBOARDS USED IN PACKAGING: CYLINDER MACHINE GRADES

Board	Uses	Chief Characteristics
Plain chipboard; solid newsboard; news vat-lined chip; filled news	Used in the manufacture of set-up boxes.	100% recycled. Lowest-cost board produced; not good for printing. Color light gray to tan. Adaptable for special lining papers.
White vat-lined chipboard	Higher-grade set-up box with white liner.	Adaptable for color printing.
Bending chip	Folding cartons for light items.	100% recycled, excellent bending. Lowest-cost board for folding cartons.
Colored manila-lined bending chip	Same as bending chip.	Bright white liner, excellent for color printing.
White-line 70 newsback	Used for folding cartons, posters, displays.	100% recycled. Smooth white board.
Clay-coated boxboard	Used for quality cartons.	Very smooth white board with excellent printing surface.
Solid manila board	All carton uses, including food.	White liner and manila back.
Extra-strength plain kraft-type board	Used for hardware, automotive and other machine parts, toys.	Recycled. Available in various colors. Excellent bending ability.
Extra-strength white-lined or clay-coated kraft-type board	Used for heavy objects requiring durability and strength.	Recycled. Top liner is white, back is brown. Available in pastel shades.

Table 2. BOXBOARDS USED IN PACKAGING: FOURDRINIER MACHINE GRADES

Board	Uses	Chief Characteristics
Uncoated solid bleached sulfate	Waxed, polycoated, or plain frozen-food cartons.	Strong white board. Hard sized for water resistance and extended service under freezer conditions. Good bending. 100% sulfate pulp.
Clay-coated solid bleached sulfate	Cartons suitable for pharmaceuticals, cosmetics, hardware.	Excellent printability, scoring, folding, and die-cutting; ovenable. Outstanding merchandising appeal. Takes all types of coatings. 100% sulfate pulp.
Clay-coated solid unbleached sulfate	Heavy-duty packages; beverage carriers; folding cartons for food, cosmetics, textiles, housewares.	Strong, moisture-resistant board. Excellent printing surface.

Table 3. PAPERS FOR PRINTING, LABELING, AND DECORATIVE PACKAGING

Type	Uses	Characteristics
Flat or dull finish, coated and uncoated	Box wraps for gifts, cosmetics, jewelry.	Smooth, excellent printing surface. Good for embossing.
Glossy finish, supercalendered	Labels, displays, box coverings.	For quality printing. Embosses well.
Cast coated	Box wraps, gift wraps, labels.	Smooth, high gloss. Brilliant white and colors. Excellent printability. Scuff resistant.
Flint	Box and gift wraps, labels for cosmetics, gifts.	Extremely fine-quality surface, high gloss. Wide color range.
Friction glaze	Box covering. Good appearance at low cost.	High glaze. Scuff resistant. Economical.
Metallic-finish Argentine	Box wraps and overwraps, luxury items. Gift packages.	Has decorative effect of foil but is less costly. Lacquer lends gold or colored look.
Foil	Decorative packaging and labels.	Aluminum laminated to paper backing. Wide range of brilliant metallic colors. Expensive.
Gravure printed	Boxes, displays, labels, platforms.	Printed with metallic powders mixed with lacquer.
Half-fine, half-fine embossed	Specialty boxes for cosmetics.	Embossed, continuous metallic surface. Wide color range. Elegant, rich.
Pyroxylin	Box coverings, food wrappers for high-quality items.	Metallic tones. Bronze, aluminum, or copper ground into pyroxylin lacquer.
Vacuum metallized	Labels for canned goods, batteries, wrappers for food and confectionary products.	Produced by vacuum metal-vapor deposit method.
Flock	Platform coverings, linings for gift boxes.	Flock of cotton adhered to surface of paper to create soft, velvety look.
Foam paper	Protective cushioning for wraps, platforms, pads, box coverings.	Foamed polystyrene sheet laminated to paper. Lustrous finish; wide range of colors.
Glitter	Box coverings, platforms.	Specks of metal embedded in paper surface.
Iridescent or pearlescent coating	Luxury items, wraps, platforms.	Mother-of-pearl effect.
Tissue	Used for jewelry, flowers, fruit, hosiery. Specially treated and designed for visual appeal.	Treated for texture, water, resistance, printability.
Glassine	Laminated trays for candies and other food products.	Grease and oil resistant.
Parchment	Wrapper, liner for greasy or oily items.	Water resistant, high wet strength. Lacquered or waxed base.
Polyethylene and saran-coated kraft	Widely used for food products. Cereal and cracker cartons, bread wraps.	Excellent barrier to moisture.

2

The Folding Carton

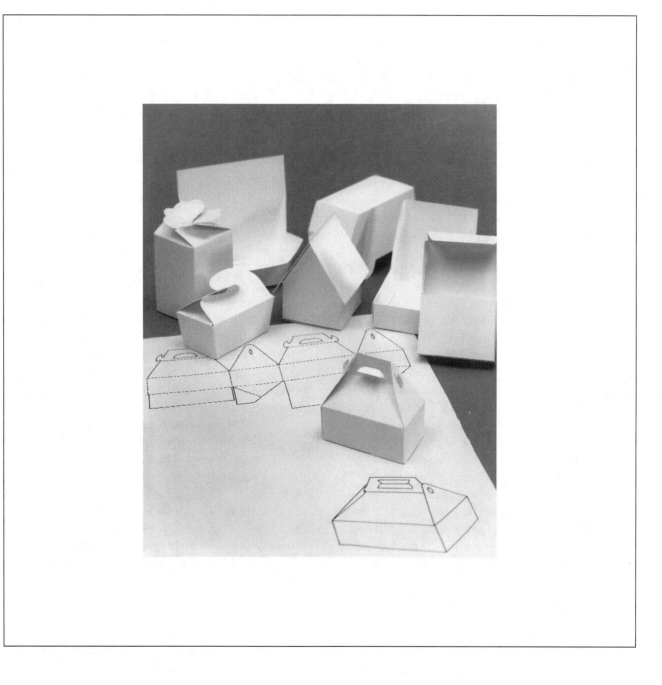

The cracker barrel and the flour sack were once the standard packages in the grocery trade. Today this method of "bulk packaging" has given way to a system in which small units are produced at manufacturing plants. The product is placed in paper containers made from a wide variety of boards.

Once in the hands of the consumer, a product may not be used all at once but only partially consumed. The package therefore must allow the consumer to remove a portion of the product without destroying the container. Dispensing of the contents over time is an important function of packages. Finding ways to protect and preserve products for future use is a primary concern of packagers.

With the advent of new methods of marketing and merchandising, packages now must be designed to be selected by the consumer without the help of a sales clerk. This development has given manufacturers and dealers an opportunity to enhance their products through the graphic and structural design of the package.

The shape, size, proportions, and material (board) of packages depend on the size and type of product they contain. To most people, the folding carton is the most familiar type of paper container. It is economical in terms of both material and production costs. Being collapsible, it takes up minimal space during shipment or storage. The versatility of the folding carton adds value and sales appeal to the product. The most advanced printing, embossing, and decorating techniques can be utilized to enhance the carton's appearance.

HISTORY OF THE FOLDING CARTON

The annals of invention are crowded with stories of fortunate accidents. Packaging is no exception. It happened in 1879 in the Brooklyn factory of Robert Gair, a successful printer and manufacturer of packaging materials. Gair was inspecting a printed seed package that had been accidentally slashed by a printing plate that had been improperly positioned on the press.

The revelation came to him: Why not create a die ruled for cutting and scoring a blank (paperboard) in a single impression? The idea worked, and the first machine-made folding carton was born.

In 1896 the first cartons of crackers appeared in stores, with a sealed inner bag to keep the product fresh. National Biscuit Co. was the first U.S. manufacturer to use this new packaging medium. The brand name was Uneeda, a name that is still in use.

By 1923 there were 200 manufacturers of folding cartons. By 1980 the number had grown to about 700 plants, which together use $3\frac{1}{2}$ million tons of paperboard annually.

There are many variations in the construction of folding cartons. The modifications that can be made in the basic pattern are limited only by the designer's imagination.

Today folding cartons are precision-made. Low-cost packages are supplied in *knock-down* form and are known as *blanks*. When assembled, they become three-dimensional rigid packages. They can be filled by high-speed automatic, semiautomatic, or hand-operated equipment. Knock-down containers lend themselves to various types of marketing and retailing systems, notably those that handle food, gifts, pharmaceuticals, cosmetics, toys, hardware, and housewares.

THE TRAY-STYLE CARTON

One of the basic types of folding carton is the *tray*. In one type of tray-style carton, a solid bottom is hinged to wide side and end walls. The sides and ends are connected by a flap, hook, locking tab, or lock. These cartons have a variety of cover and flap parts extending from the walls and sides of the tray.

In another type of tray carton, two pieces, one slightly smaller than the other, form the base and cover of a two-piece telescoping box. Typical tray packages include cigarette cartons, bakery trays, ice cream cartons, pizza cartons, and garment carriers.

THE TUBE-STYLE CARTON

Another basic type of folding carton is the *tube*. The body of the tube-style carton is a sheet of board that is folded over and glued against its edges to form a rectangular sleeve. It has openings on the top and bottom that are closed with flaps, reverse or straight tucks, and locks.

Tube-style cartons give the product fully enclosed protection. They are used for bottled products, cosmetics, and pharmaceuticals. Often a window is added so that the consumer can see the product. Many unusual tube types and styles are available, including contoured, triangular, octagonal, and even rounded shapes. The following pages present a variety of folding-carton patterns. They can be constructed in any size from board or sheet plastic (acetate).

COSMETICS, PHARMACEUTICALS, AND PROMOTIONAL PACKAGING

The cosmetics industry was the first to effectively use packaging, and is still inspiration for the most creative types of packaging.

The history of cosmetics goes back thousands of years. The Bible describes the practice of anointing the head and body with oil. Primitive tribal symbols applied to the face and body were used to frighten enemies in various cultures in Africa, North and South America, and Oceania. Another traditional use of cosmetics is to attract members of the opposite sex. (The most famous figure associated with this practice is Cleopatra, the last queen of Egypt, who was renowned for her skill in making and applying cosmetics.) The ancient Greeks and Romans were great users of all types of cosmetics— hair dyes, creams, makeup, and the most important toiletry, soap, which was invented around 100 A.D. In the seventeenth century Queen Elizabeth I used dyes (red) for her hair as well as fragrances; some of the most well-known historical figures of this period never bathed, but used perfumes and rosewater to counteract body odors and dirt.

The wholesale manufacturing of cosmetics and toiletries began in France during the reign of the "Sun King," Louis XIV (1643–1713). During the Napoleonic era in the early 1800s, it became a major industry.

Cosmetics packages have the look of luxury and elegance. The standard cosmetics container is a folding or set-up carton embellished with rich colors, textures, and graphics. It attempts to portray the consumer as potentially beautiful, desirable, and elegant. From the marketing point of view, the cosmetics package is selling hope.

Pharmaceuticals are another important area for the packaging industry. There are about 1,000 drug-manufacturing companies in the United States. Pharmaceuticals packaging uses all types of packaging materials, including paper, board, and plastics.

Over-the-counter drugs must have sales appeal, and packaging plays an important role in providing that appeal. The folding carton encloses a glass or plastic container for the medication. The carton provides protection as well as serving as a promotional tool.

Pharmaceuticals packaging involves several factors, including moisture and oxygen control to preserve the potency of the medication; volatility and light protection to protect the drug against exposure to harmful ultraviolet rays; and protection against heat, which can cause deterioration of the product. Safety also plays a significant role. Many items (e.g., surgical dressings, injectables) must be kept sterile until they are used. The packaging must also be childproof and tamperproof.

The requirements for packaging of cosmetics and pharmaceuticals are very similar. In both cases manufacture and distribution is strictly regulated by the Food and Drug Administration.

Promotional packaging was originally used by the pharmaceuticals industry for the introduction and promotion of new products. The packages were sent to physicians. Gradually these packages became a total-communication effort. In addition to product samples, the package contained descriptive literature, reply cards, and videotapes. Manufacturers of other products and services soon followed the lead of the pharmaceuticals industry, and promotional packaging is now used by all major industries. Examples of mailers and other imaginative promotional packages are shown on the following pages.

UNUSUAL APPROACHES TO CARTON AND BOX CONSTRUCTION

Designers today are increasingly interested in technology. Computer-aided design, for example, enables them to create or alter concepts in minutes. Such technological advances have reduced the need for handwork and special skills. However, student and professional designers alike have a tendency to cling to familiar methods and skills, which give them a feeling of self-fulfillment.

Like new technologies, economic factors appear to limit the possibilities for design, yet they can actually stimulate innovation. An example is the revitalization of folding-carton design that occurred as a result of the prohibitive cost of set-up boxes and the increasing cost of hand labor.

Many of the following pages present examples of structural experiments. They reflect the exciting experimental work being done in schools of design. Innovations in design like those shown here can strongly influence the marketing of a product, often surpassing even the expectations of the client.

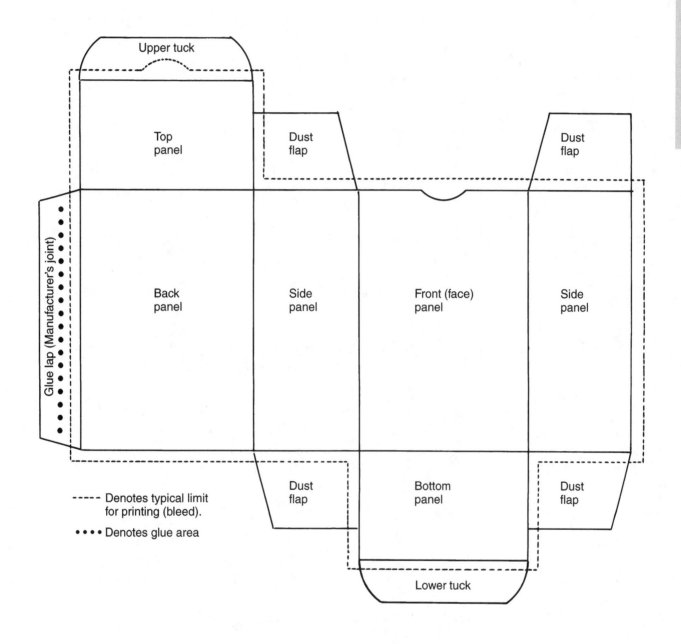

Upper tuck

Top panel

Dust flap

Dust flap

Glue lap (Manufacturer's joint)

Back panel

Side panel

Front (face) panel

Side panel

Dust flap

Bottom panel

Dust flap

Lower tuck

---- Denotes typical limit for printing (bleed).

•••• Denotes glue area

Trays

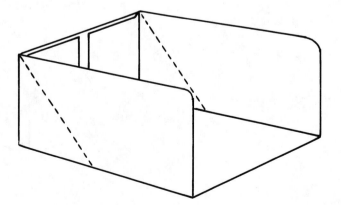

OPEN ENDED COLLAPSIBLE TRAY, VARIATION

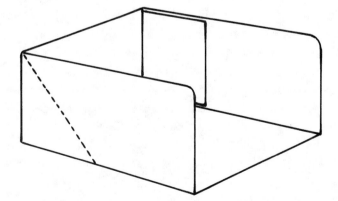

GLUED-UP TRAY WITH PROTECTIVE END PANELS

**GLUED-UP TRAY WITH PROTECTIVE END PANELS AND TAPERED
SIDE PANELS**

The Folding Carton

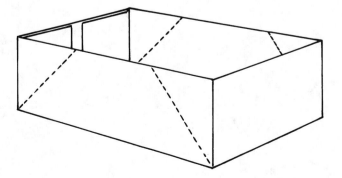

FOUR CORNER BRIGHTWOOD STYLE TRAY

BEERS STYLE TRAY WITH RAISED END PANELS

The Folding Carton

STRIPPER LOCK TRAY

GLUE OR HEAT-SEALED TAPERED TRAY WITH FLANGE

ONE PIECE BRIGHTWOOD TRAY WITH HINGED COVER
AND DUSTFLAPS

DOUBLE WALL TRAY WITH DOUBLE WALL LID

The Folding Carton

The Folding Carton

The Folding Carton

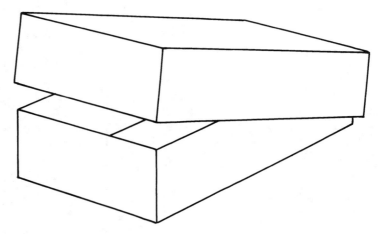

TWO PIECE FULL TELESCOPING HARDWARE STYLE TRAY
WITH LOCK ENDS

SEAL END STYLE (EASY OPEN) TRAY

The Folding Carton

Ideal for a small tissue dispenser that fits in pocket-book or briefcase.

The Folding Carton

SINGLE END WALL, DOUBLE SIDE WALL TRAY

PARTIAL DOUBLE END WALL AND DOUBLE SIDE WALL TRAY

FOOT-LOCK DOUBLE WALL TRAY

ARTHUR LOCK OR JONES LOCK TRAY

DOUBLE END WALL, SINGLE SIDE WALL TRAY

DOUBLE END WALL, SINGLE SIDE WALL TRAY

SIMPLEX STYLE TRAY, WEB CORNER SPOT GLUED

FRAME VIEW TRAY (VARIATION)

This tray was used for crafts projects and features an enclosed compartment for small parts.

FRAME VIEW TRAY (VARIATION)

FRAME VIEW TRAY WITH LID WITH LOCK TABS

The Folding Carton

Combined with the overwrap pattern on the following page, this package lends itself to numerous graphic possibilities.

This wrap fits around the view tray of the preceding page; the four flaps fit neatly inside the tray.

Grain Direction

The Folding Carton

The Folding Carton

**KLIKLOK "CHARLOTTE" STRIPPER LOCK TRAY WITH
HINGED COVER AND ANCHOR LOCK AND SIDELOCKS**

SIMPLEX STYLE VARIATION WITH TAPERED OUTSIDE END AND SIDE WALLS

**TAPERED SIDEWALL WITH ARTHUR LOCKS AND DOUBLE
HOLLOW SIDE WALLS**

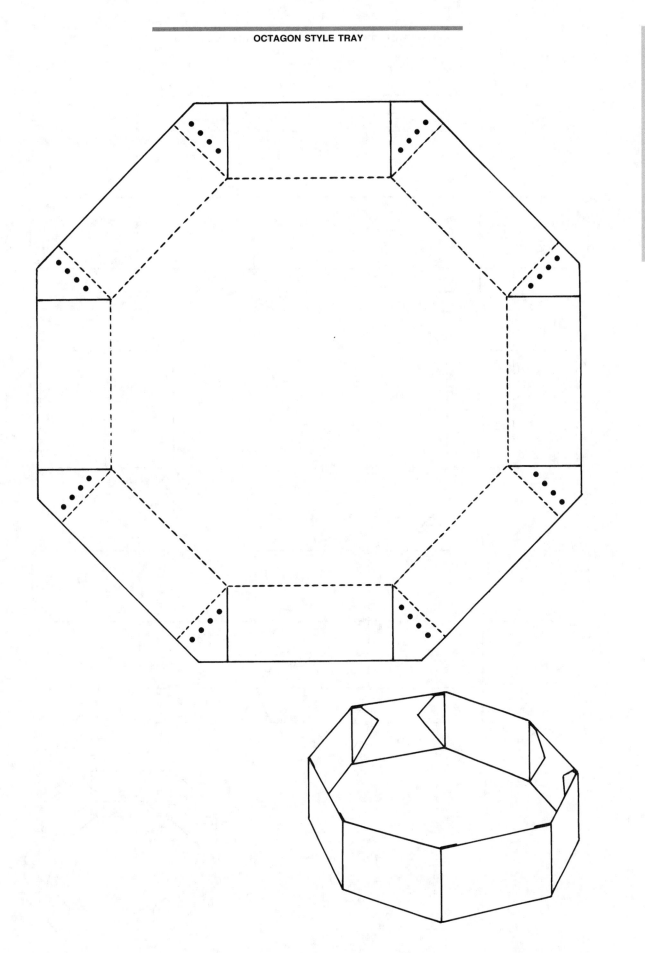

DOUBLE WALLED OCTAGON STYLE TRAY WITH FOOT LOCKS

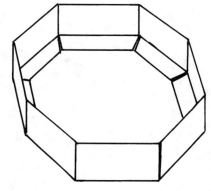

**SIX-SIDED DOUBLE WALLED TRAY AND PARTIAL
TELESCOPING COVER**

The Folding Carton

This specially constructed tray has a tapered interior that makes it an excellent container for pens and similar items.

This variation of the preceding pattern is adaptable for multiunit packaging.

Two Simplex style trays with the partition on the following page make an interesting design for a four-compartment tray.

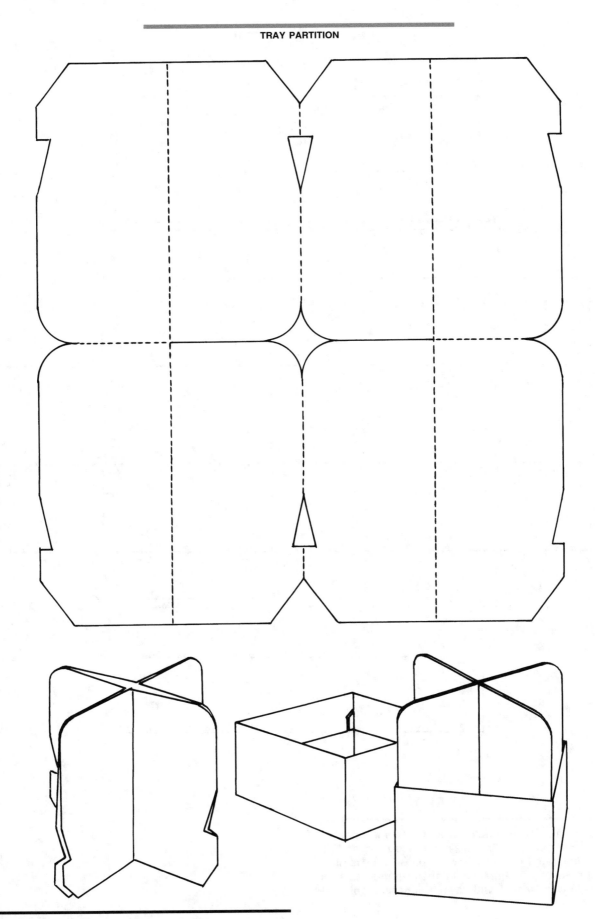

The notches in the lower part of the partition will lock
into the partial double side walls of the tray.

The Folding Carton

This variation of the Simplex style tray has two double walled cover sections that meet in the top center. The insert illustrated on the following page indicates the many possibilities inherent in these patterns. This package was sealed with a self-adhesive gold-foil sticker.

The insert may be die-cut and scored in any configuration desired.

The Folding Carton

A more complex example of a frame view tray.

This one-piece scored and die-cut tray may be used for eggs, fruit, candy, and the like.

Tubes

FULL OVERLAP SEAL END WITH PRE-PERFORATED POUR SPOUT

The Folding Carton

The Folding Carton

BELLOW TUCK TOP, HIMES LOCK AUTOMATIC BOTTOM
(ALSO KNOWN AS CRASH LOCK)

STRAIGHT TUCK WITH DRUGSTORE LOCK

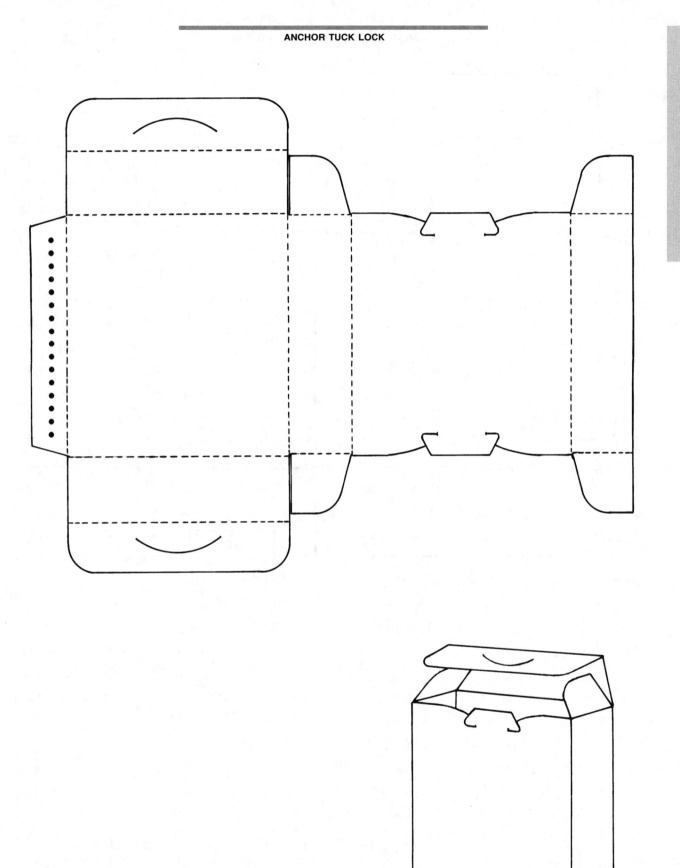

FRICTION TUCK TOP, SNAP-LOCK (OR HOUGHLAND) BOTTOM

The Folding Carton

QUAD LOCK-EAR HOOK DOUBLE LOCK BOTTOM

**STRAIGHT TUCK WITH FULL OVERLAPPING FLAPS
AND SIDE HOOKS**

A typical butter carton.

The Folding Carton

MARGARINE/BUTTER CARTON WITH TEAR-AWAY TABS

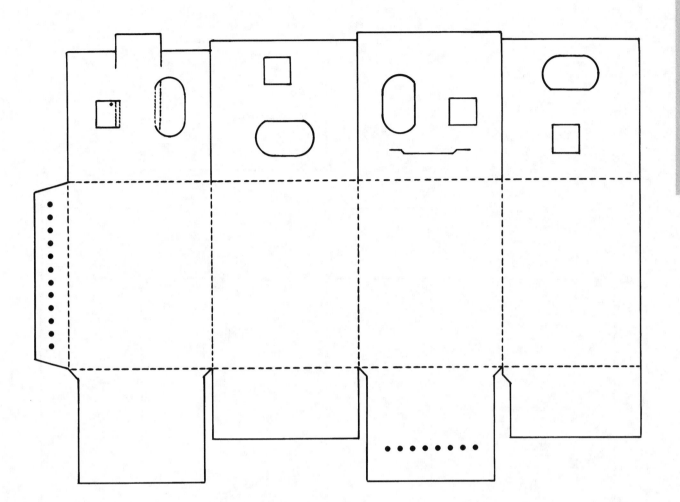

Especially useful for heavy loads.

The Folding Carton

SEAL END WITH LOCK TAB AND ZIPPER

A tissue dispenser.

A tissue dispenser.

The Folding Carton

Grain Direction

The double fifth panel permits printing on one side of the board, with copy appearing on the back panel.

The Folding Carton

Grain Direction

Specially designed for lightbulbs.

The Folding Carton

SIDE-BY-SIDE REVERSE TUCK/STRAIGHT TUCK CARTONS,
VARYING IN HEIGHT

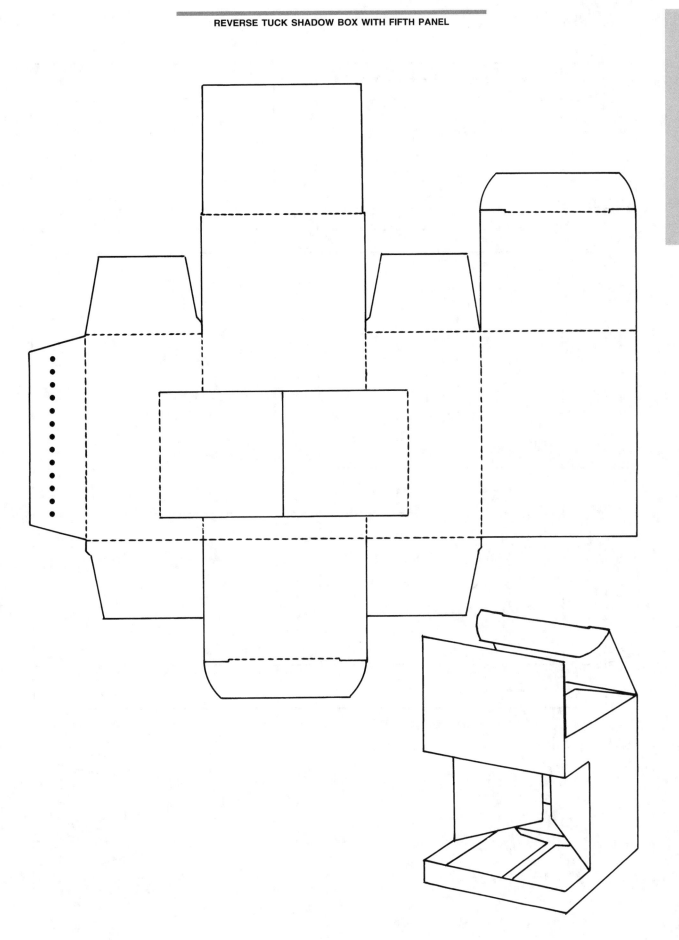

ONE PIECE OPEN VIEW DISPLAY CARTON

**PACKAGE FOR A PHARMACEUTICAL PRODUCT, WITH
HANGING PANEL**

DOUBLE COVER WINDOW BOX

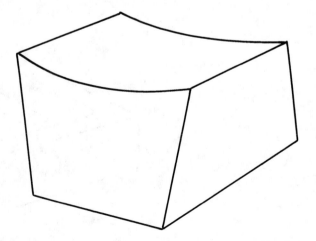

The special effect is created by the concave top and convex front or face panel.

Alternating closure creates an interesting graphic effect that may be enhanced with color.

The curved scores on this pattern create a concave top
and bottom and convex front, side, and back panels.
The tension of the scores keep both top and bottom
closed.

Specialty closure on top panels gives free rein to many graphic solutions.

Specially shaped top closure creates an interesting concave/convex interplay.

The Folding Carton

Scored front, back, and side panels lend themselves to cord or ribbon decoration.

**FULL-FLAP AUTOMATIC BOTTOM, TAPERED TOP WITH LOCK
TABS AND CARRYING HANDLE**

TAPERED SIDES, 1-2-3 BOTTOM TONGUE LOCK, TOP PANEL WITH CENTERED CARRYING HANDLE

WAX-COATED LIQUID (MILK) CONTAINER

The typical milk carton lends itself to many different applications. *(©Excello)*

One variation of the milk carton style container has a
slit/loop closure.

The curvilinear scores in the four riser panels serve as a unique decorative alternative.

A different treatment of the top side-panel scores and
closure tab. This carton features an auto-lock bottom.

Manipulating the panel by means of die cuts creates many interesting design variations.

In combination with the patterns for the inserts on the facing page, this package with carrying handle was created for six small body lotion bottles.

SPECIALLY SCORED TOP AND BOTTOM WITH FULL OVERLAP
LOCK TAB AND SLIT

A carton with a variation of Arthur locks and 1-2-3 bottom develops into an attractive alternative packaging form.

Ideal for ribbons, tape, etc.

The Folding Carton

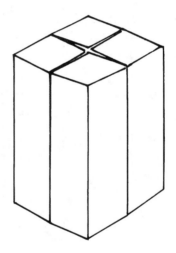

This quadruple box is ideal for sets of items such as cosmetics.

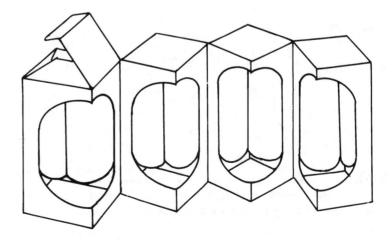

ONE PIECE QUADRUPLE BOX

The Folding Carton

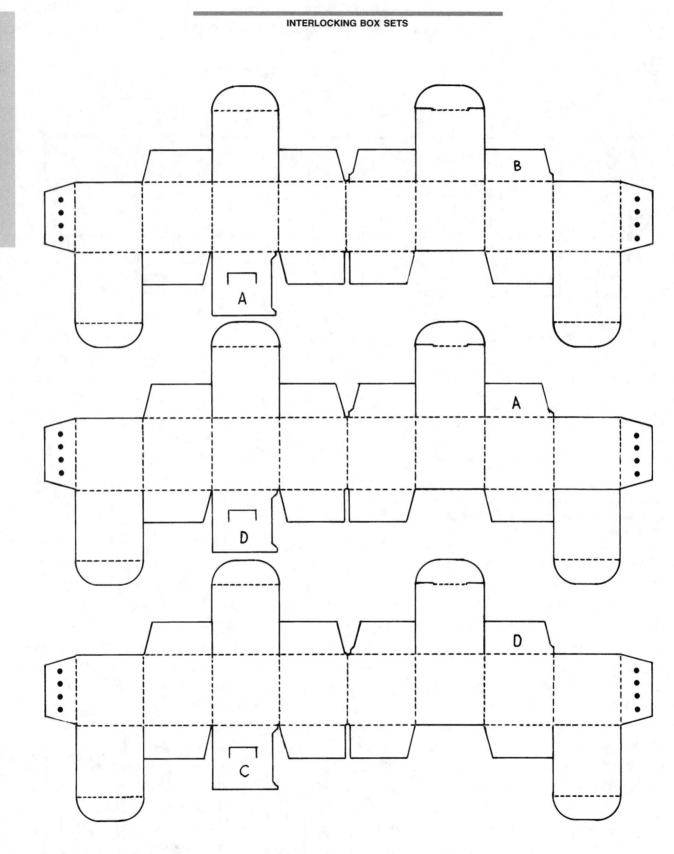

**The eight cubes interlock at the dust flaps as indicated.
This allows for many graphic innovations.**

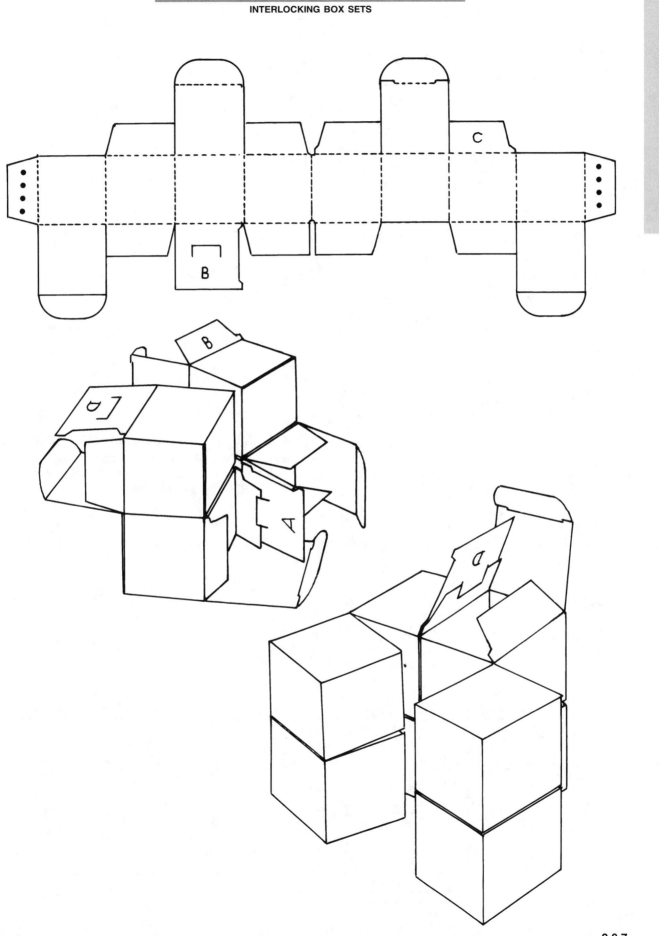

The Folding Carton

The Folding Carton

Grain Direction

This one-piece shadow box, designed by Yick-Man Liu, lends itself to a great assortment of graphics and makes a beautiful gift package.

This is but one of the many design possibilities for top closure.

This package opens when the sleeve at the top is pushed down.

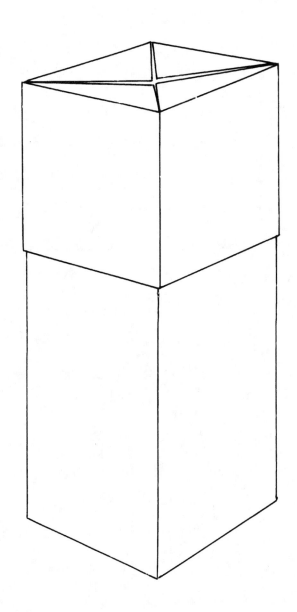

As the top overlapping sleeve is pushed down, the triangular closure shapes open. Sliding the sleeve up will close the top.

Curved scores create many interesting effects, which
may be enhanced by innovative printing.

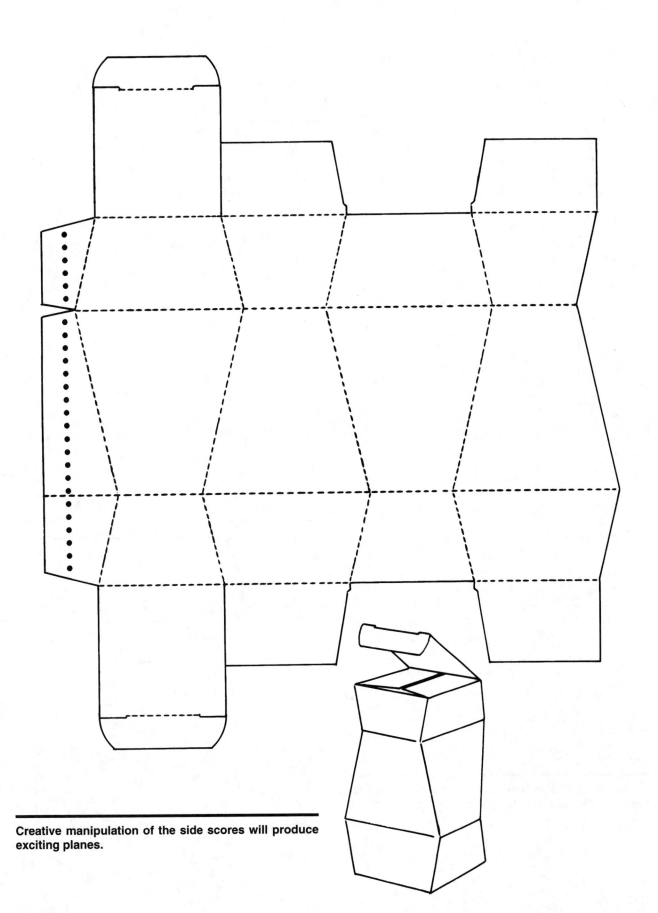

Creative manipulation of the side scores will produce exciting planes.

This straight tuck carton, with its concave front and back panels, has many design possibilities.

The many curves on this carton allow for a beautiful interplay of shadows.

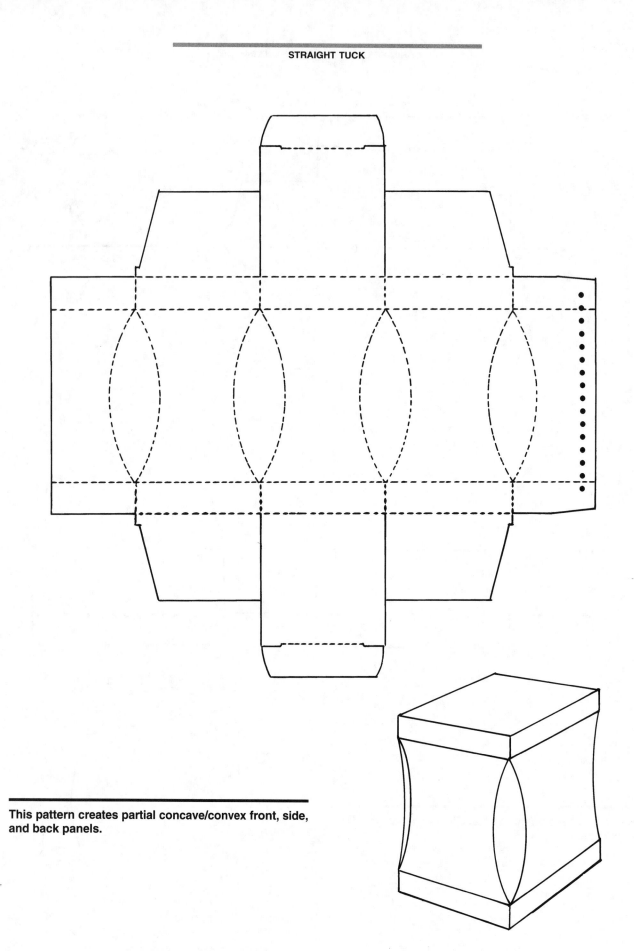

This pattern creates partial concave/convex front, side, and back panels.

**TRIANGLE WITH TUCK CLOSURES AND TEAR-AWAY
ZIPPER STRIP**

The scores in the panels create a fifth riser panel on this carton.

The side and top panels of this carton allow for great graphic visibility.

The additional score would be needed in order to ship the carton flat.

SIX-SIDED CARTON WITH STRAIGHT TUCKS
(PAT #3,148,443)

An alternate version of a six-sided package.

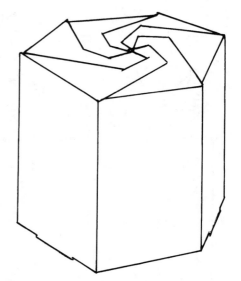

The small triangles have been die-cut out of the top closure to permit easier closing.

**SIX-SIDED CARTON WITH PUSH-IN CLOSURE AND AUTOMATIC
GLUED BOTTOM**

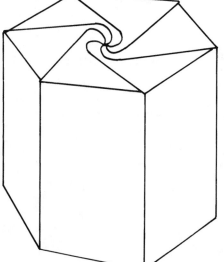

A variation of the previous two patterns, this carton also has an auto-glued bottom.

A variation of the 1-2-3 bottom.

SIX-SIDED DISPENSER PACKAGE WITH TEAR-AWAY OPENING
(Pat. Kliklok)

This carton has an intriguing shape created by the scoring of the riser panels.

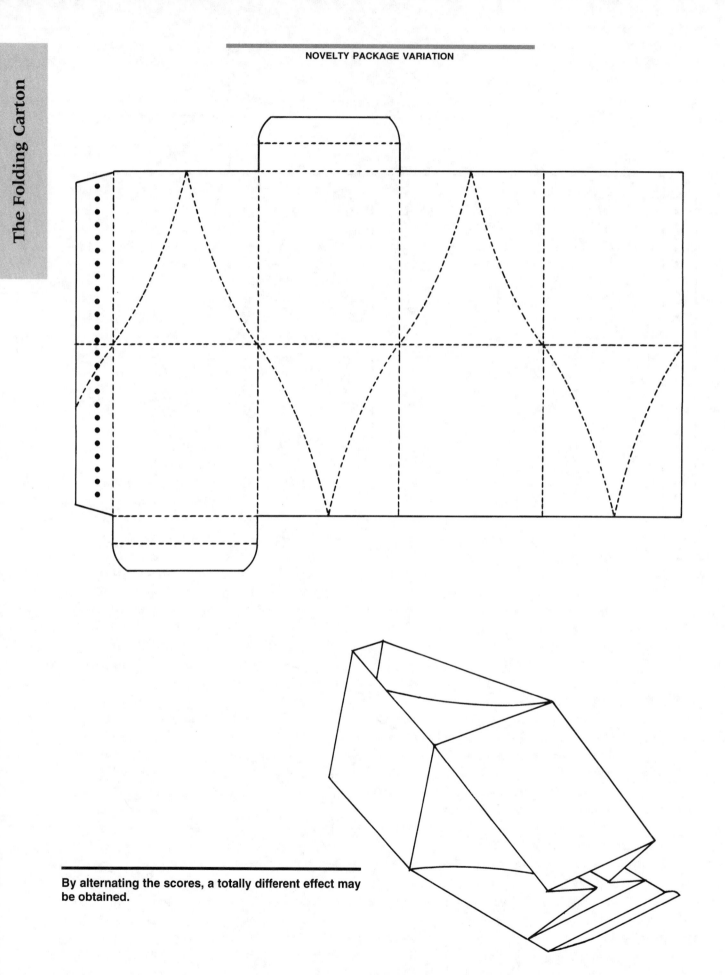

By alternating the scores, a totally different effect may
be obtained.

TEARDROP SHAPE OPEN CONTAINER

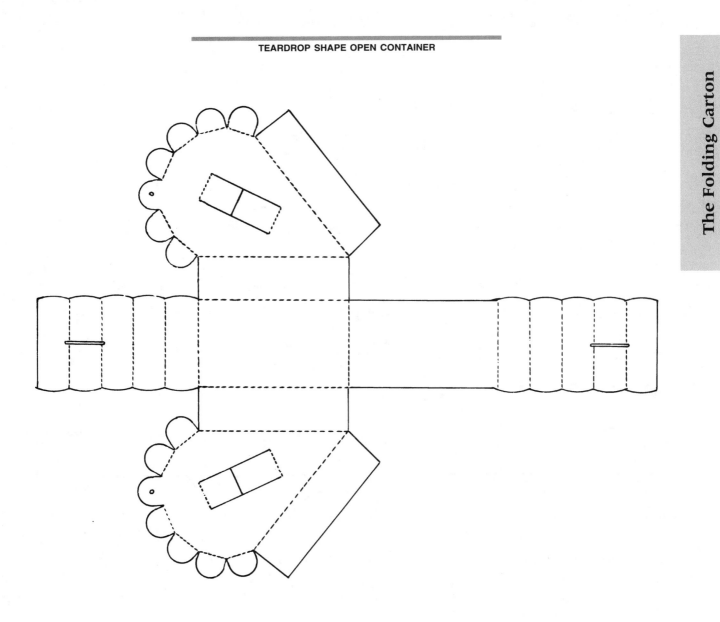

This and the following patterns may be executed in many different stocks. With die-cut shapes to match their products, they make ideal holiday gifts/ornaments.

This house-shaped package only needs to be decorated to fit the specific needs of the product.

The Folding Carton

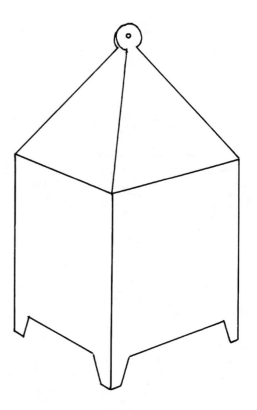

Lock tab and slit options make this package totally glue free.

This package also requires no glue.

A container that depicts a traditional Dutch Burgher house makes an attractive keepsake.

A somewhat more complex pattern could lead to a series of keepsakes.

The Folding Carton

This package lends itself to recreations of famous buildings by means of creative die-cutting of windows and graphics.

An alternative approach to the pattern on the preceding page.

This little flower stall is an ideal package for cosmetics.

The shape of this box speaks for itself. An inserted pouch will make up for the holes created by the die-cut wheels.

SIX-WHEELER TRUCK

This novelty item requires no glue.

This old-fashioned sleigh has graphic application possibilities.

This accordion is held together with the strap provided.

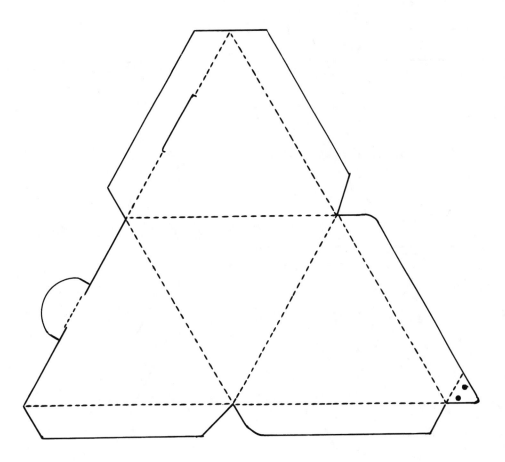

Spot gluing of the flap off the lid helps keep the package closed.

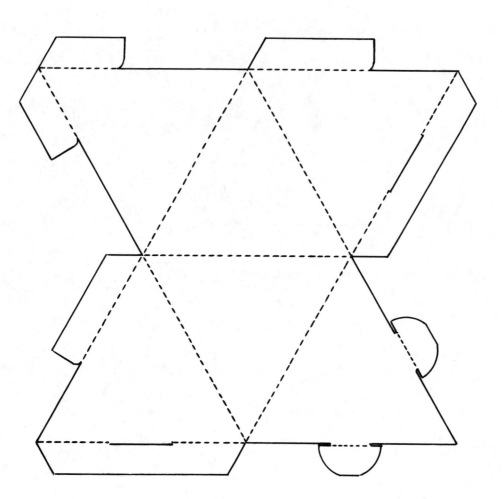

The flaps off the lid lock into the shortened dustflaps.

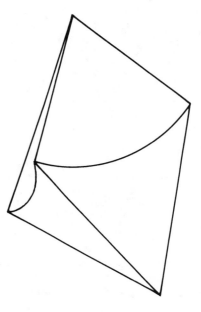

Straight and curved scores permit a wide range of variations.

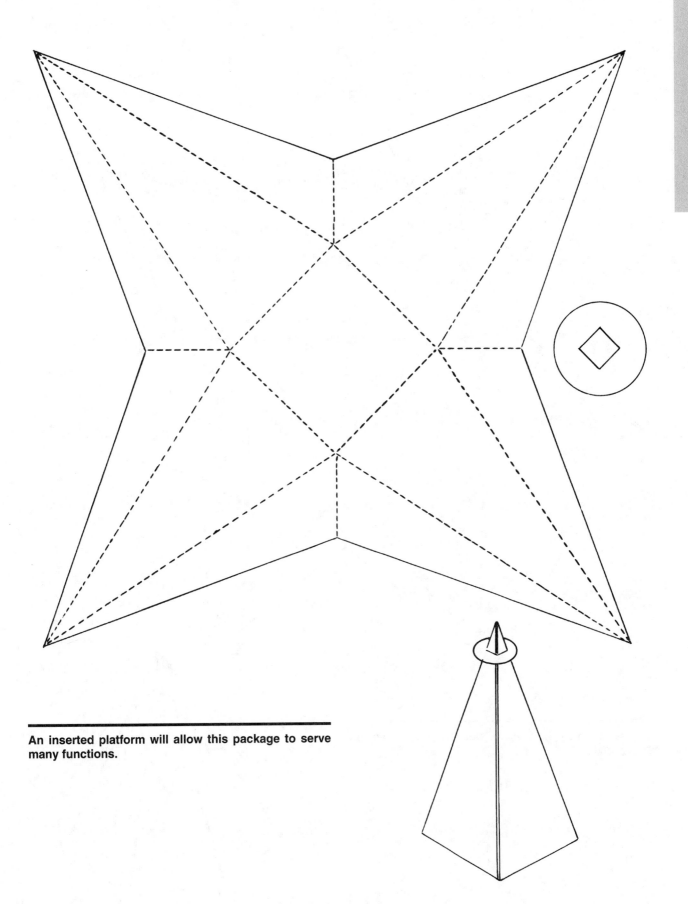

An inserted platform will allow this package to serve many functions.

This exciting package uses an elasticized string as a
closing device.

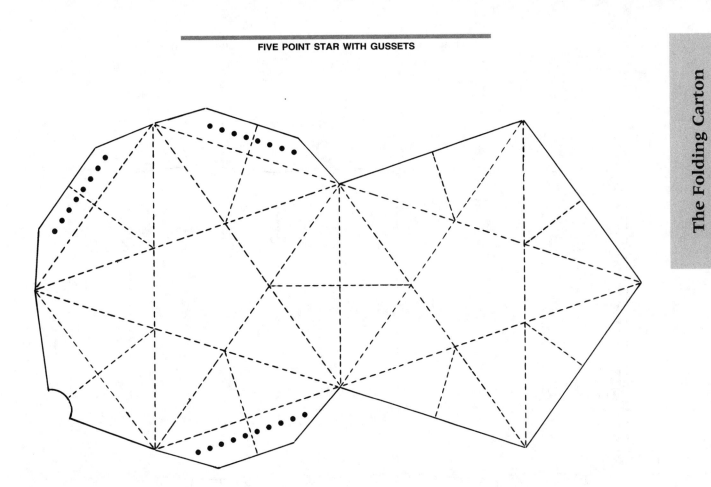

This beautiful candy package could be used as a holiday ornament.

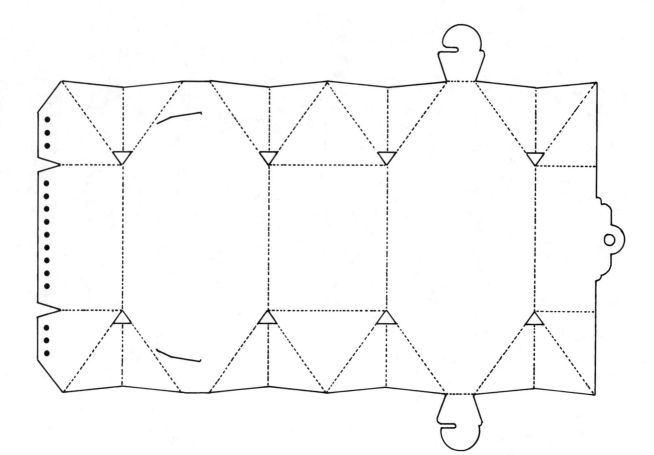

Designed by David James & Associates, this tea package lends itself to a variety of applications.

Sleeves, Wraps, and Folders

Typically used for videotape cassettes.

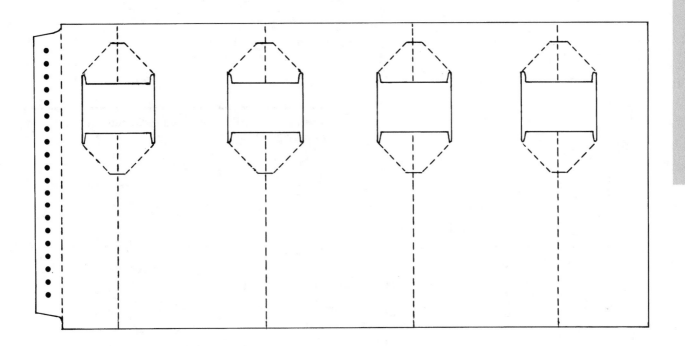

The Folding Carton

Typical construction for a fragile product (e.g., light-bulbs). *(PAT. #3,082,931)*

Grain Direction

The tension of the inner panel die-cuts help keep the product in place.

LIGHTBULB SLEEVE WITH SCORED INFOLDS
AND PARTIAL COVER

Both the cover and the infolds may be constructed to fit
any product.

May be adapted for many other product uses.
(PAT. #3,630,430)

This one-piece construction lends itself to a large variety of lightweight product packaging with maximum product visibility.

Grain Direction

The Folding Carton

BOTTOM LOCK SLEEVE

Used for scarves, hosiery, etc.

The Folding Carton

May be spot glued.

**SHALLOW SIDE WALL FOLDER WITH WINDOW FOR
PRODUCT IDENTIFICATION**

Perfect for hosiery.

The lid will snap under the front dust flap.

Typically used for photographic prints and negatives.
The scored gusset allows for expansion.

The Folding Carton

Grain Direction

The Greenleaf style closure makes this sleeve easy to open for inspection of the product. A window may be die-cut in the package.

The Folding Carton

This package is developed from two interlocking sleeves with a carrying handle.

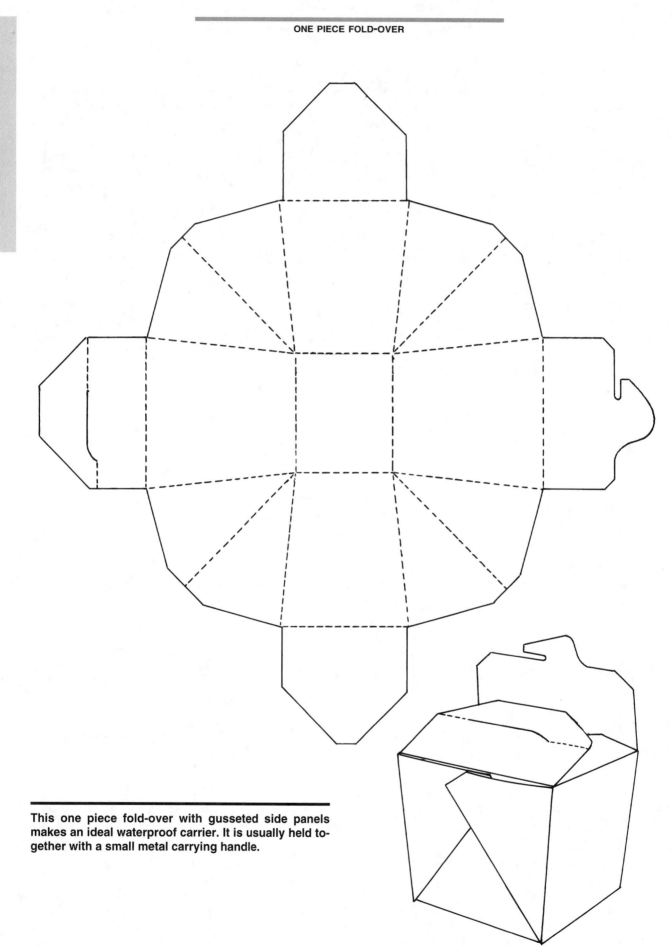

This one piece fold-over with gusseted side panels makes an ideal waterproof carrier. It is usually held together with a small metal carrying handle.

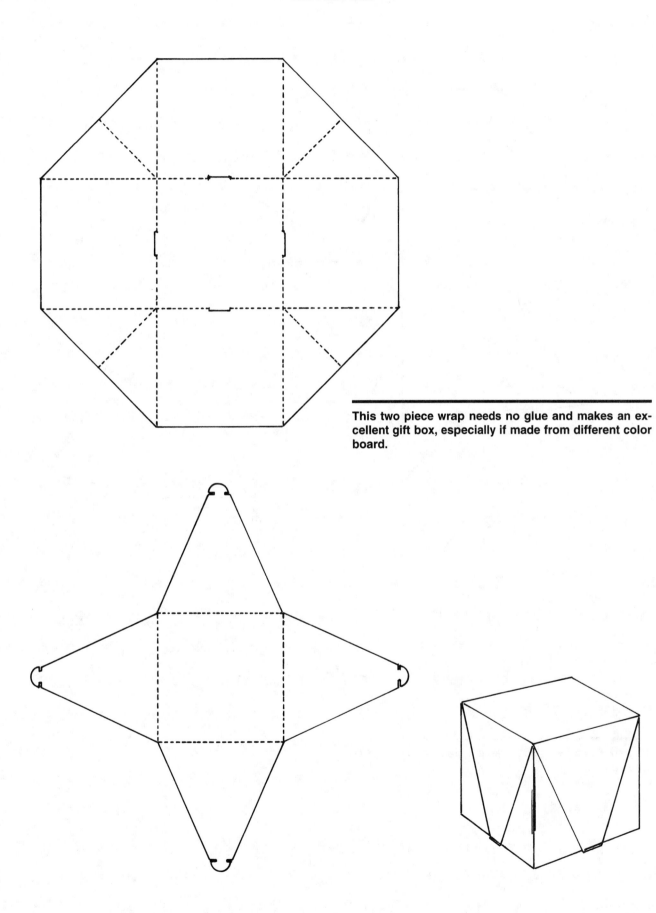

This two piece wrap needs no glue and makes an excellent gift box, especially if made from different color board.

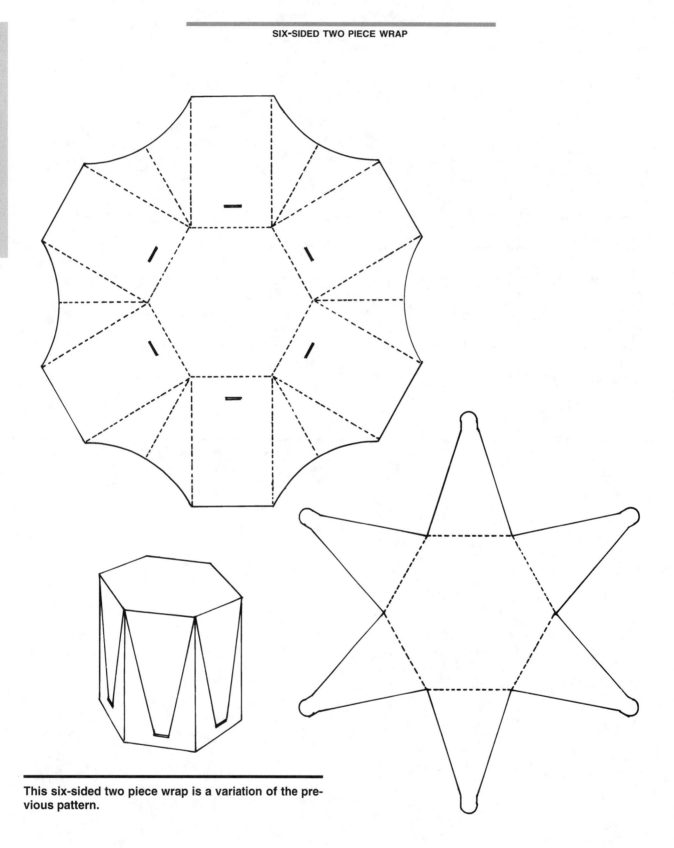

This six-sided two piece wrap is a variation of the pre-vious pattern.

The lid with wings holds the four sides in place.

Adding four fold-over flaps on the riser panels as in the next pattern, will serve as a locking device for the lid.

The Folding Carton

Folding the gussets to the exterior of the package creates an interesting visual effect.

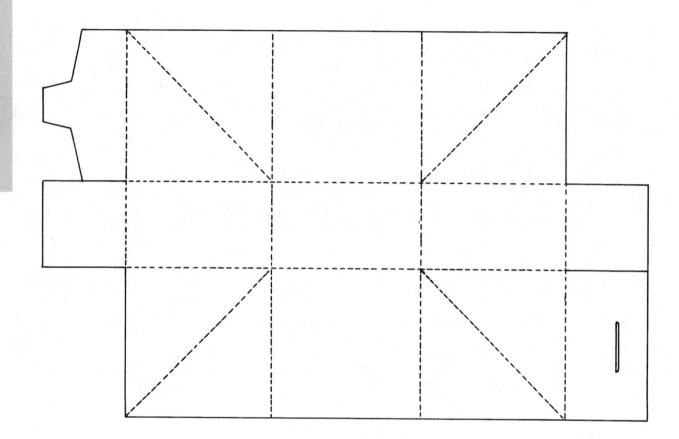

This wrap makes a handsome container, without the use of adhesives.

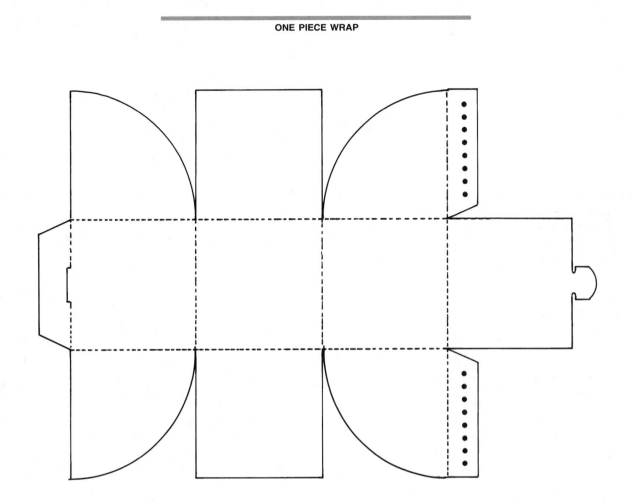

A one piece wrap that shows an interest-catching configuration on the sides.

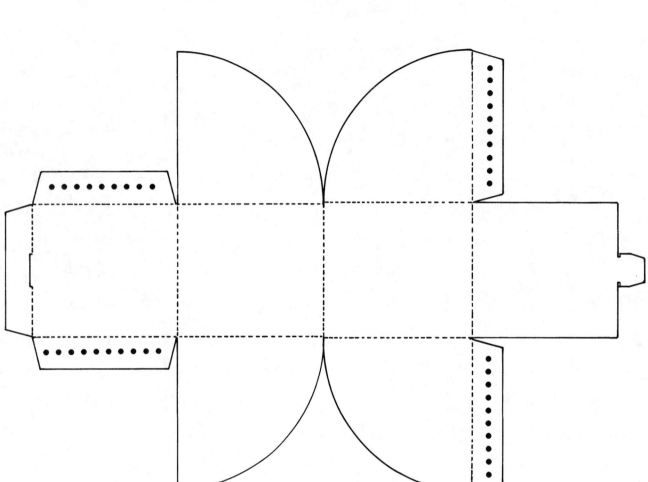

A rearrangement of the panels and glue flaps alters the appearance.

This fold-over wrap has been partially scored in order to create a soft, rounded top when assembled.

Suitable for cards as well as haberdashery items.

When taped shut, it makes an excellent mailer.

This box with Arthur locks and hook locks needs no glue. It makes an excellent container for jewelry items.

FOLDER WITH HOOKS FOR STRING OR ELASTIC-BAND CLOSURE

A simple, economical, one-piece compact disc folder with good product protection, easy product access, and sufficient printing surface for product information.

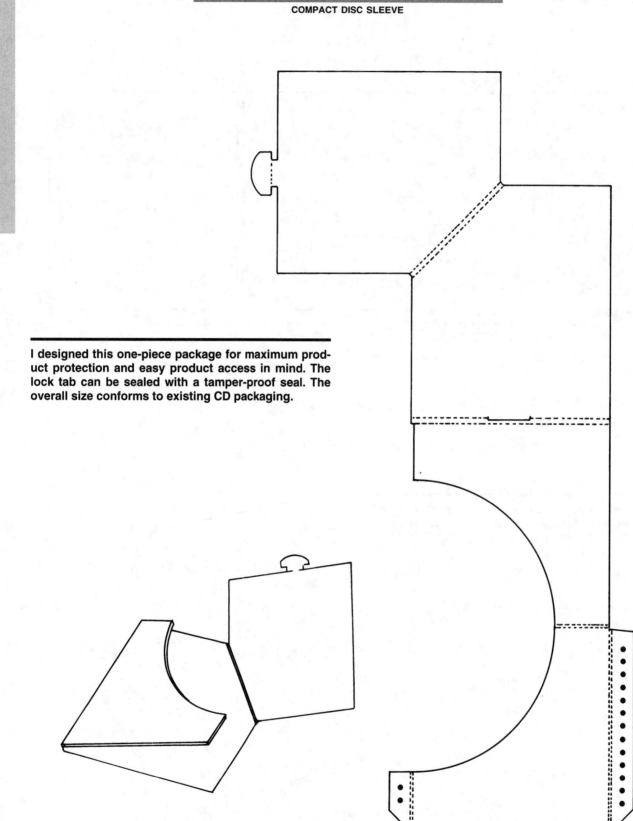

I designed this one-piece package for maximum product protection and easy product access in mind. The lock tab can be sealed with a tamper-proof seal. The overall size conforms to existing CD packaging.

Two piece 2 CD sleeve based on the old vinyl gramo-
phone record packaging.

The Folding Carton

This GW compact disc folder pops up when the outer
sleeve is opened. The CD may be accessed by its spin-
dle hole, thus avoiding contact with grooves.

The package is formed and sterilized around the product.

TETRAHEDRON FOR ASEPTIC PACKAGING OF LIQUIDS

Also suitable for novelty items, hosiery, etc.

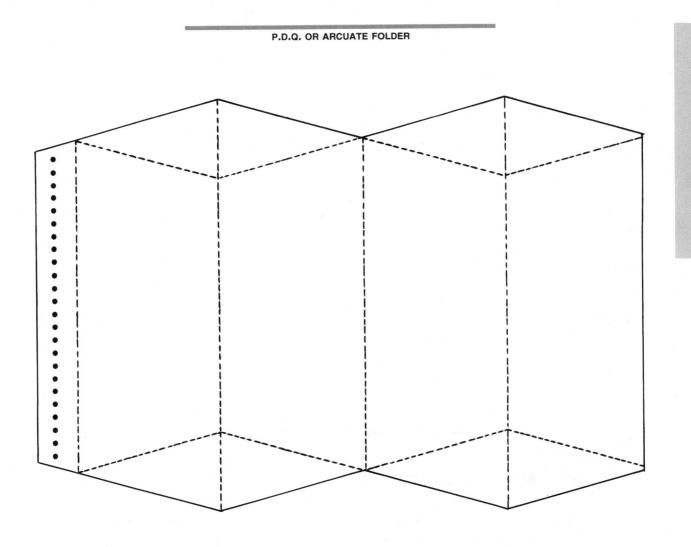

The end closures snap in.

PILLOW PACK

Grain Direction

This Pillow Pack features a cushioned compartment.

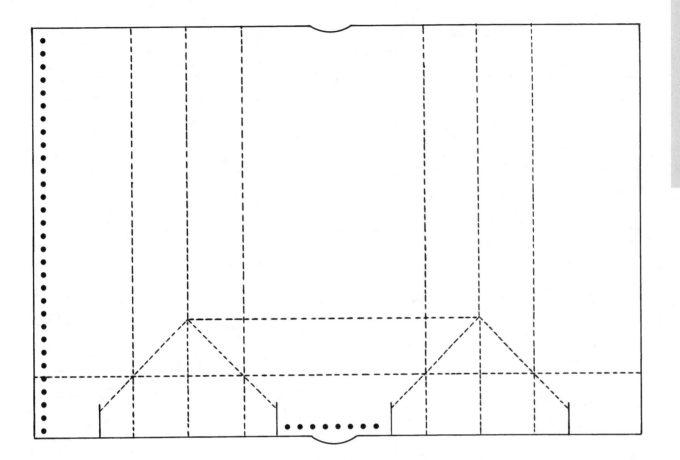

Typical prototype for grocery bag, shopping bag, etc.

**BAG WITH GREENLEAF BOTTOM, SIDE GUSSETS, AND HOOK
LOCK CLOSURE**

Grain Direction

GUSSETED BAG WITH HIMES LOCK AUTOMATIC BOTTOM,
GUSSETED SIDES, AND FLAPS ON COVER THAT
LOCK INTO GUSSET

POUCH WITH HIMES LOCK AUTOMATIC BOTTOM, SIDE GUSSETS, AND REINFORCED SNAP-LOCK CLOSURE

An adaptation of the standard office file produces a
neat package for soft goods.

**BAG WITH HIMES LOCK AUTOMATIC BOTTOM, GUSSETED SIDES,
AND CARRYING HANDLE WITH SNAP-LOCK CLOSURE**

This carrying satchel with double Anchor lock bottom
and lock-tab/slit top is one of many possible variations.

This triangular carry-all needs no glue and may be used for many different gift items.

**TWO PIECE, SINGLE LAP CONTAINER WITH OUTSIDE ROLLED
TOP AND INSIDE FOLDED BOTTOM**

**TWO PIECE, SINGLE LAP CONTAINER LID WITH
INSIDE ROLLED TOP**

**TWO PIECE, SINGLE LAP, TAPERED CUP WITH INSIDE FOLDED
BOTTOM AND OUTSIDE ROLLED TOP**

The Folding Carton

Used predominantly as the core for rolls of paper and mailing tubes.

Spiral-wound tubes may be used as the basis for circular packaging.

Cosmetics and toiletries often make use of this form. The following page shows the use of a tube as the core of a novelty package.

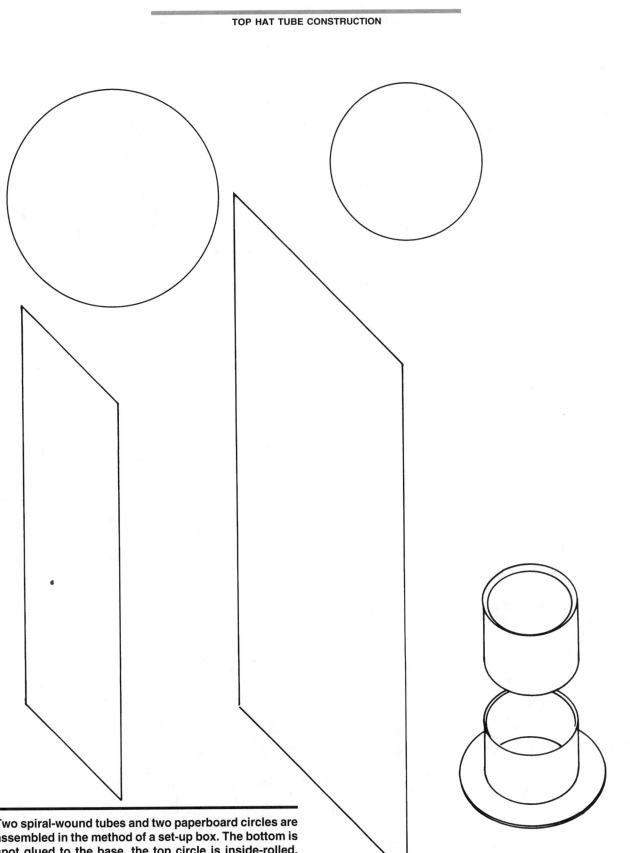

Two spiral-wound tubes and two paperboard circles are assembled in the method of a set-up box. The bottom is spot glued to the base, the top circle is inside-rolled. The slightly larger tube slides over the bottom to create a top-hat.

SLEEVE STYLE THREE-COUNT MULTIPACK ACE LOCK BOTTOM

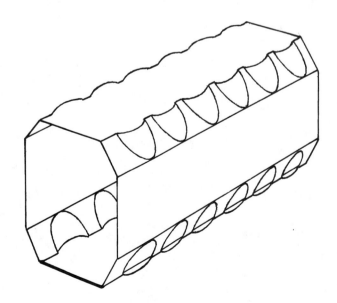

ANDRE-MATIC CAN CARRIER WITH HEMMED EDGE

**END LOADED BOTTLE CARRIER WITH TAPERED SIDES,
REINFORCED HANDLE, AND SEPARATE INSERTS**

TWELVE-COUNT CAN CARRIER WITH PRE-PERFORATED
TEAR-OPEN STRIP
(PAT. #3,955,740)

**SIX BOTTLE WING LOCKED SLEEVE WITH TOP AND BOTTOM
CHIME LOCKS**

When glued up and assembled, the handle will consist
of four layers of board, the center divider of two layers,
and the outside and bottom of a single layer.

The Folding Carton

3

The Set-Up or Rigid Paper Box

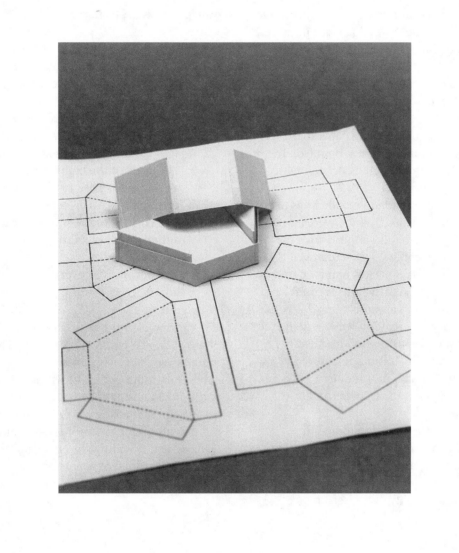

The word *box* is derived from the boxwood baskets made in ancient Rome. By the thirteenth century, the Chinese were making boxes from decorated cardboard made of rice chaff, in which they exported tea. In France, attractive boxes were made for hats, garments, cosmetics, and jewelry in the seventeenth and eighteenth centuries. These boxes were covered with handprinted decorative papers and fabrics. During the nineteenth century, attractive boxes for toys and games were produced in Germany and England.

The first American boxes using paperboard were made in Boston by Colonel Andrew Dennison in 1844. The name Dennison is still associated with many kinds of paper products.

Store clerks often used crude wooden jigs to form cardboard into "cartons," which were closed and tied with string. From this simple device the set-up box evolved. According to historians, it is possible that an artistic salesperson hand lettered the store's name on the box. This event could be considered the birth of the brand name.

The set-up paper box is produced and delivered as a three-dimensional construction ready to be packed with merchandise. It is manufactured in a series of simple steps that do not require costly dies or machinery. Production runs can be small, medium, or large.

The basic set-up box can be modified to accommodate unusual requirements; transparent plastic domes and platforms, a variety of hinges, embossing, gold stamping, lids, and compartments can be added. The versatility of the set-up box is valuable because it adds to the display and gift appeal of boxed items. It is no accident that set-up boxes are so widely used in the merchandising of cosmetics, jewelry, candy, and gifts. The box not only provides protection but has great promotional value.

Setup-up boxes are made from four basic materials: boxboard, corner stays, adhesives, and covering materials. The process is as follows:

1. Flat sheets of boxboard are cut and scored to size. Individual blanks are separated from the sheet and the corners are cut out.
2. The sides of the blank are folded at right angles to the base to form the sides of the tray.
3. The corners of the tray are sealed by adhering paper, fabric, or metal to reinforce the corners.
4. The decorative covering is adhered to the box.

Variations in the basic shape and dimensions of boxes can be made in order to produce an infinite range of designs. Possible shapes include cubes, pyramids, cones, hearts, and stars. Boxes with hinged lids, telescoping boxes, boxes with compartments, boxes with slide-in drawers, and boxes with plastic thermoformed platforms can also be designed. Coverings or wraps may be selected from a wide range of papers or fabrics, which provide both protection and decorative effect.

Board

Cover stock

Inside cover stock

A single board is covered with cover stock on the outside, with edges overlapping the inside. The inside is covered with cover stock slightly smaller than the overall dimension.

Two pieces of board are attached with hinging tape, then covered with cover stock.

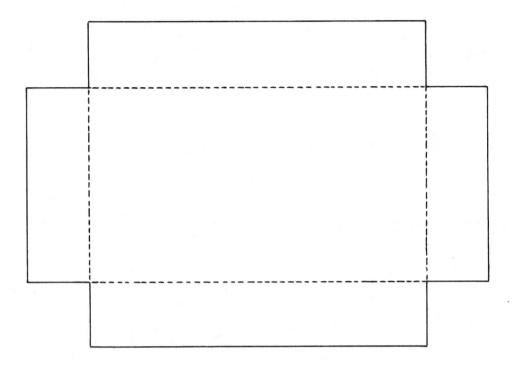

1. The stock is scored with cutting dies that cut approximately halfway through the board. The corners are then cut away so that only the bottom and side panels remain. The side panels are bent up and fastened at the corners with stay tape.

2. The assembled box is placed in a precut, preglued sheet of cover stock, which is folded around the outside and partially into the inside of the box.

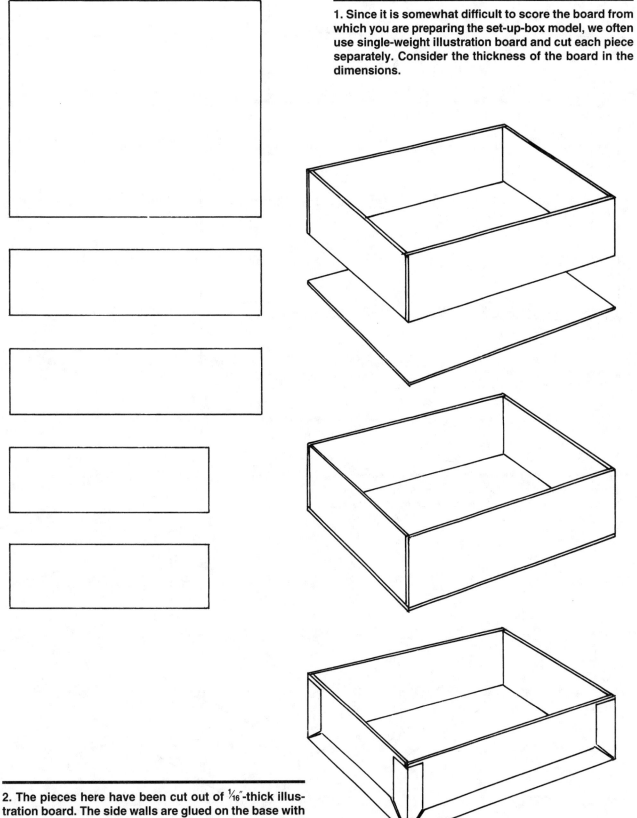

1. Since it is somewhat difficult to score the board from which you are preparing the set-up-box model, we often use single-weight illustration board and cut each piece separately. Consider the thickness of the board in the dimensions.

2. The pieces here have been cut out of $\frac{1}{16}''$-thick illustration board. The side walls are glued on the base with white glue; then the end pieces are fitted in. White masking tape is used in lieu of stay tape.

The Set-Up or Rigid Paper Box

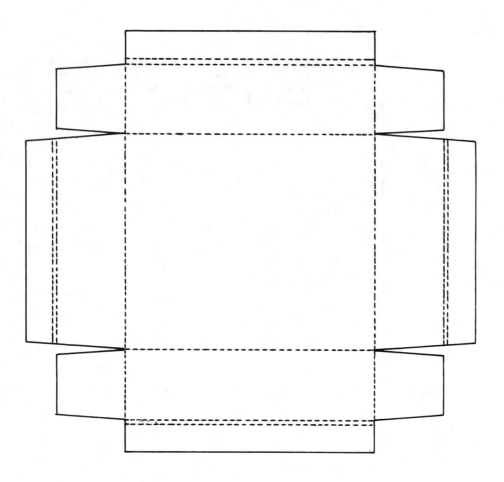

3. The assembly is covered with a nice cover stock. For this you may use rubber cement or, preferably, bookbinder's glue.

4. Next we build a platform with an oval cutout. This platform is measured to fit snugly into the tray we have just completed. When measuring, consider also the thickness of the cover stock and adjust the size of the platform accordingly.

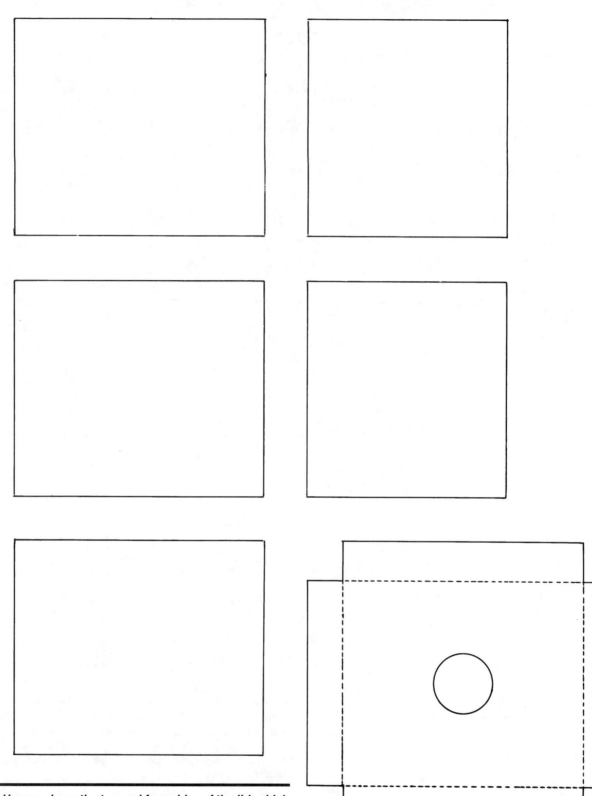

5. Here we have the top and four sides of the lid, which should have the same width and depth as the base, since it will fit over the platform.

6. This insert is for the lid and need not be covered with cover stock.

The Set-Up or Rigid Paper Box

7. Glue the insert into the lid; the circle was cut to fit the cap of the bottle to be packaged.

8. The platform may be glued into the base; the result is a handsome rigid or set-up box. If you wish a small part of the platform to show, you need only place an additional liner inside the cover. This will prevent the cover from sliding all the way down to the base.

The Set-Up or Rigid Paper Box

379

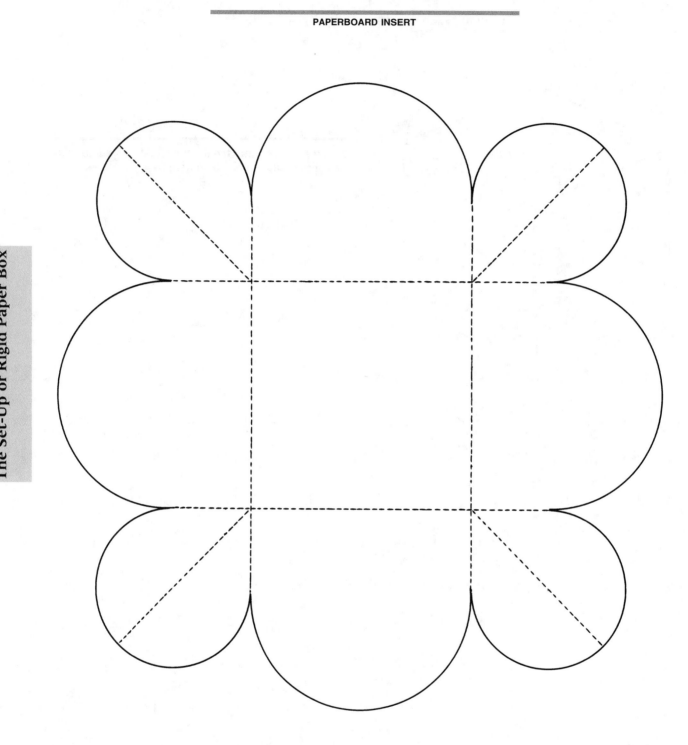

The Set-Up or Rigid Paper Box

This is a folded piece of boxboard to be used as an insert for the following rigid box.

Here the folding insert is glued onto the lined base.

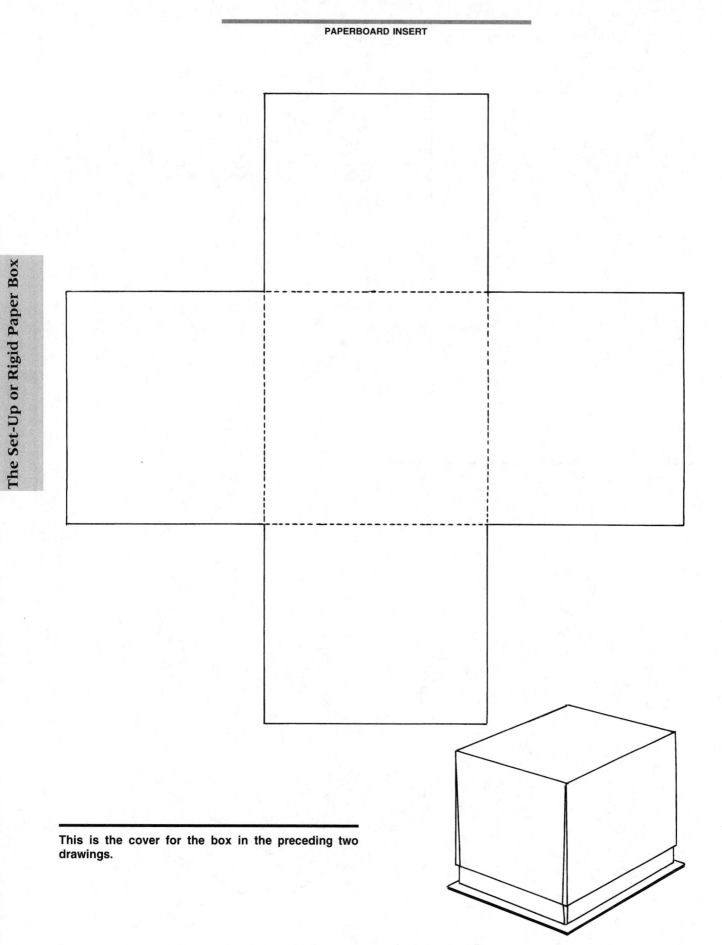

This is the cover for the box in the preceding two drawings.

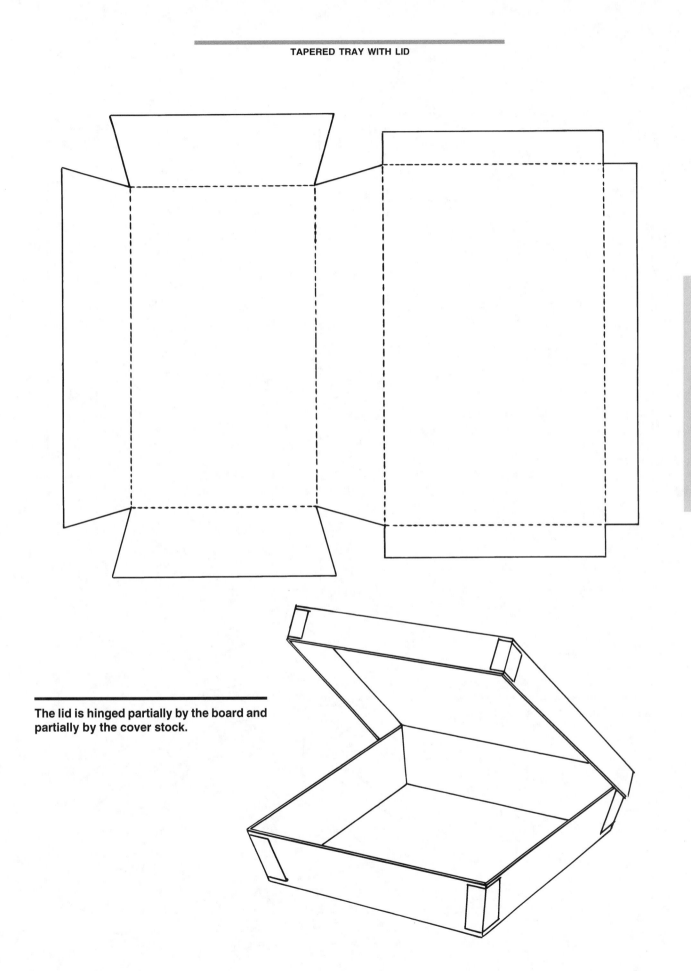

The Set-Up or Rigid Paper Box

The lid is hinged partially by the board and partially by the cover stock.

This is the pattern for the cover stock of the tapered tray with lid.

The Set-Up or Rigid Paper Box

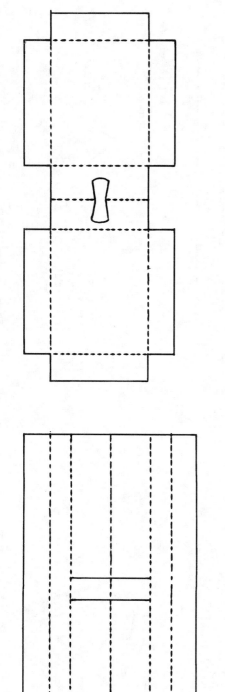

These are inserts for the tray illustrated on the preceding page. The top would be a typical insert for a souvenir spoon. The bottom insert would be for an item such as a pen.

A platform for the box illustrated on the preceding page. This would be ideal for a necklace.

A platform for necklace and earrings.

Slanted platforms for jewelry such as a tie pin or stick-pin.

This box has side panels that are slightly higher than the front and back. Once the lid has been attached with hinging tape and/or the cover stock, the front of the lid will rest on the top of the front panel.

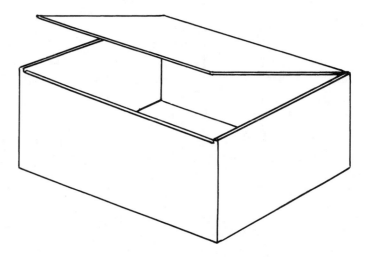

The Set-Up or Rigid Paper Box

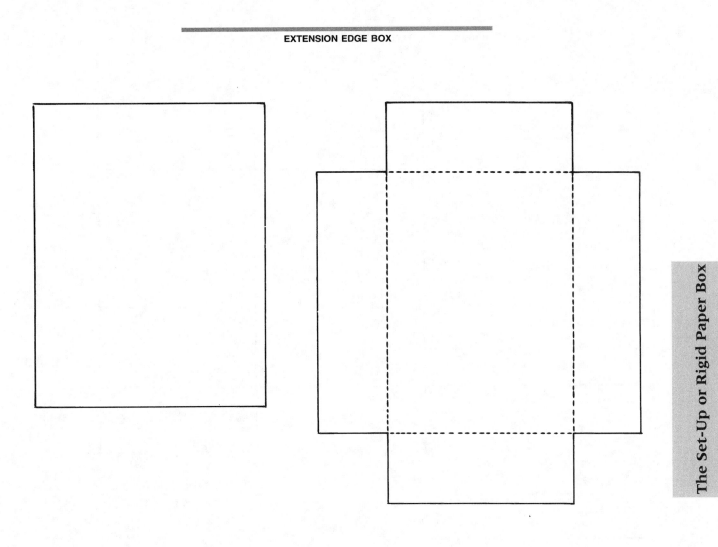

The tray is assembled and glued to the platform. Then the assembly is covered with cover stock.

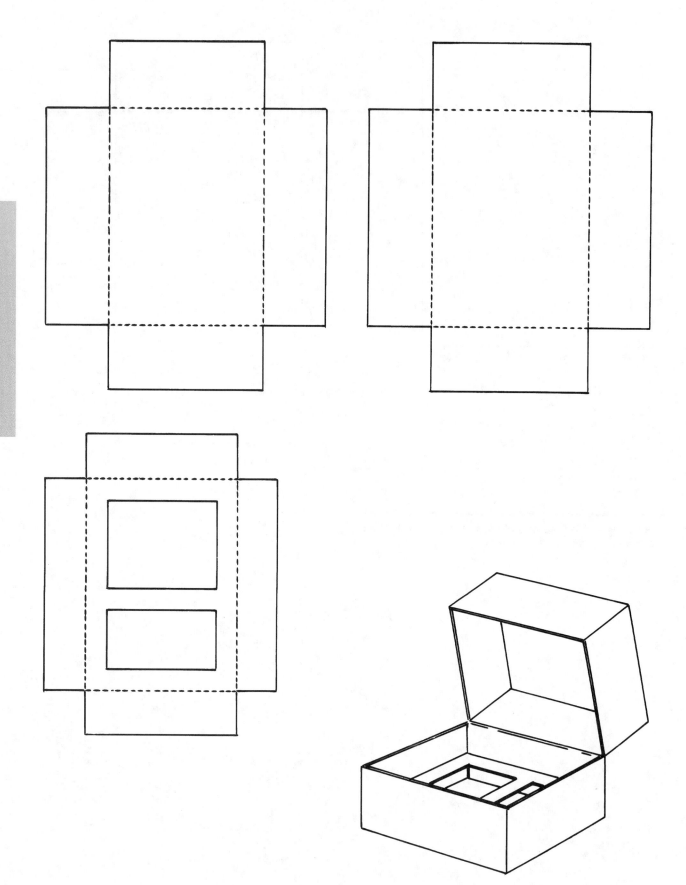

The Set-Up or Rigid Paper Box

The platform may be die-cut to fit the shape to be packaged.

The three pieces that form the outside of the box have been fastened together with hinging tape, then covered.

This box simulates a book; the spine may be shaped in order to carry the illusion even further.

The Set-Up or Rigid Paper Box

Set-up boxes may be manufactured in many shapes. This triangular box with raised platform is only one of the many geometric possibilities.

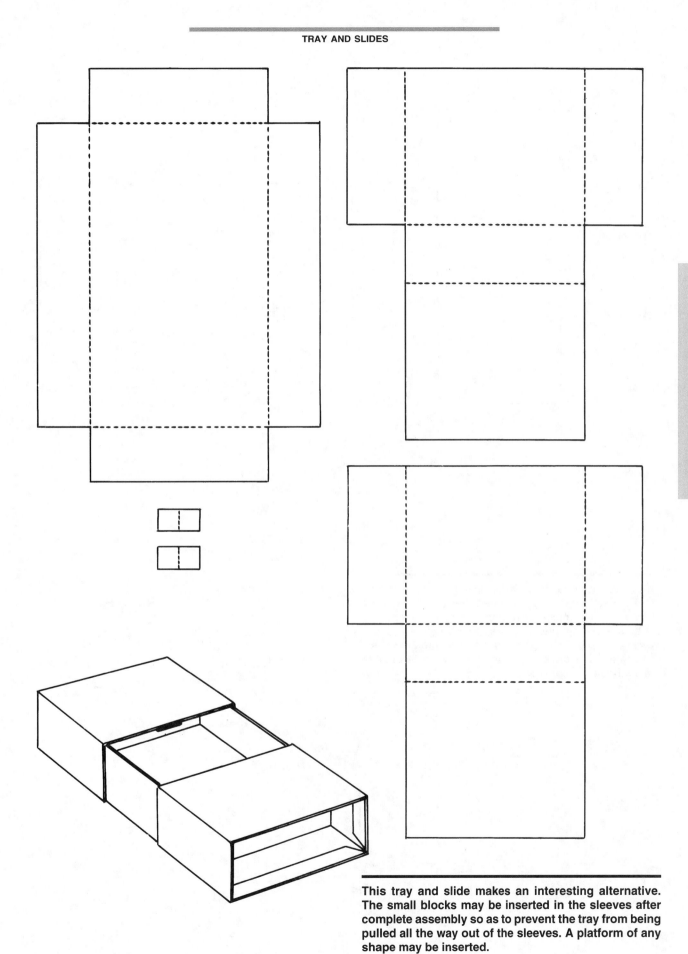

This tray and slide makes an interesting alternative. The small blocks may be inserted in the sleeves after complete assembly so as to prevent the tray from being pulled all the way out of the sleeves. A platform of any shape may be inserted.

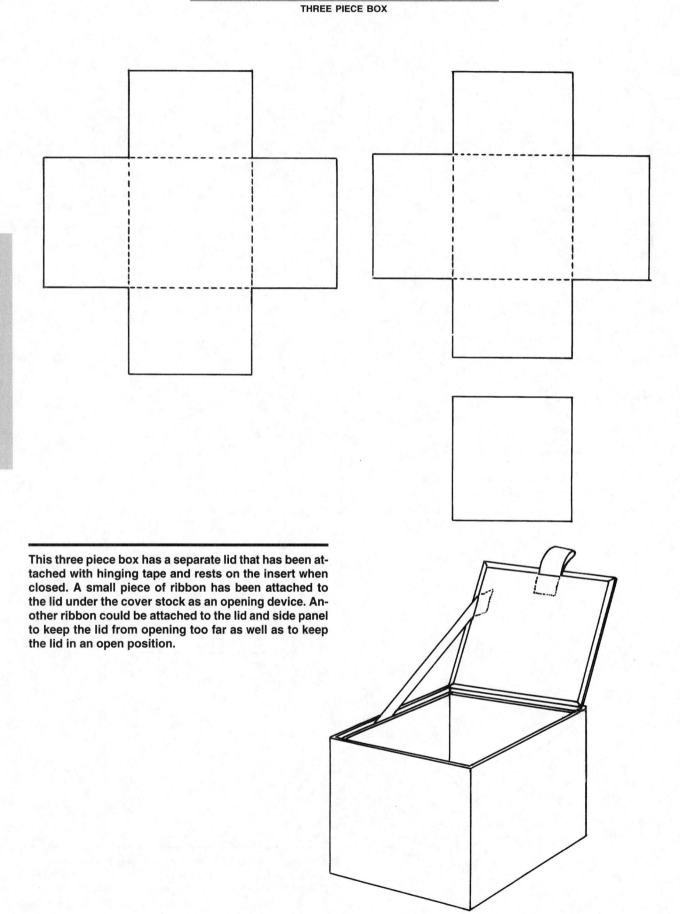

This three piece box has a separate lid that has been attached with hinging tape and rests on the insert when closed. A small piece of ribbon has been attached to the lid under the cover stock as an opening device. Another ribbon could be attached to the lid and side panel to keep the lid from opening too far as well as to keep the lid in an open position.

This file box may be used for documents, diskettes, or, with the proper insert platform, cosmetic items.

The Set-Up or Rigid Paper Box

The first stages for a four drawer cabinet are the outer sleeve and four inner sleeves to act as drawer guides.

Upon assembly, the face of the drawer guides may be covered with cover stock. Then the outer sleeve is covered with either the same or a contrasting cover.

The Set-Up or Rigid Paper Box

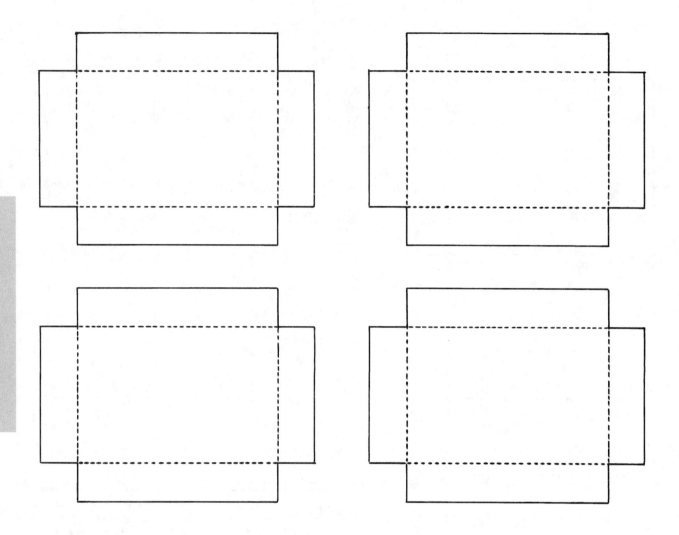

The drawers, when assembled, should be covered and
checked for a snug fit. Any available hardware may be
used as drawer pulls.

The Set-Up or Rigid Paper Box

The first stage of this box, which has eleven separate components, is the assembly of the box, lid, and front panel lid. The cover and lid are attached with hinging tape, then covered with cover stock.

The Set-Up or Rigid Paper Box

Subsequent steps are insertion of a die-cut platform into the lid, attachment of a slightly higher inner sleeve that allows room for the front to close flush with the sides, and attachment of an additional piece to the front cover that permits it to be locked in place when the lid is closed over it.

Finally, the base platform is assembled and inserted into the bottom of the box.

The Set-Up or Rigid Paper Box

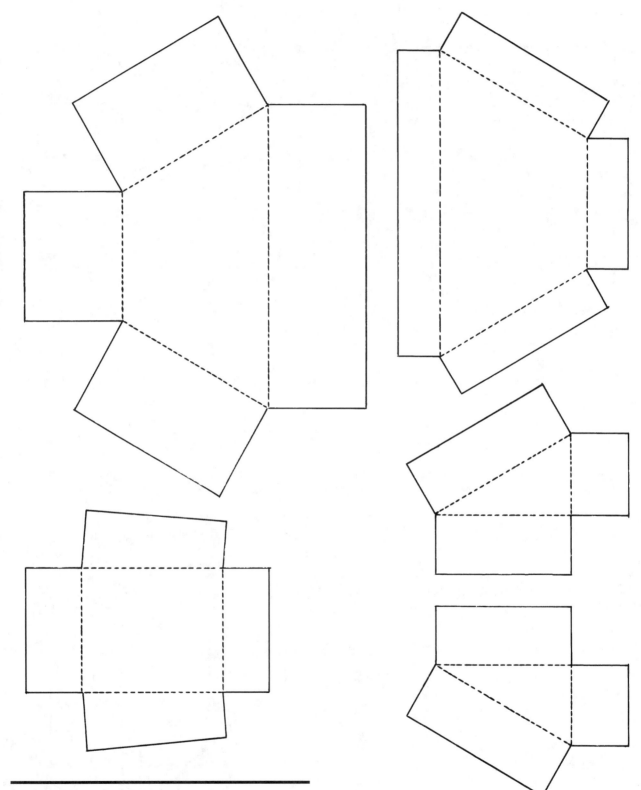

Many varieties of platforms may be assembled in order to create visual excitement.

As long as the platforms have been made to the correct size, they may be inserted in various directions. In this example the front of the set-up box may be oriented either way, and the lid may be attached accordingly.

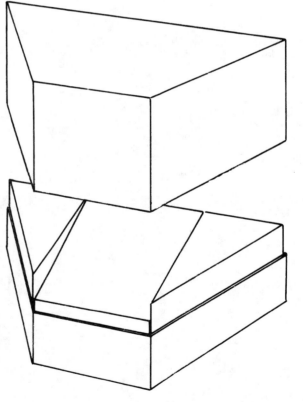

4

Corrugated Containers

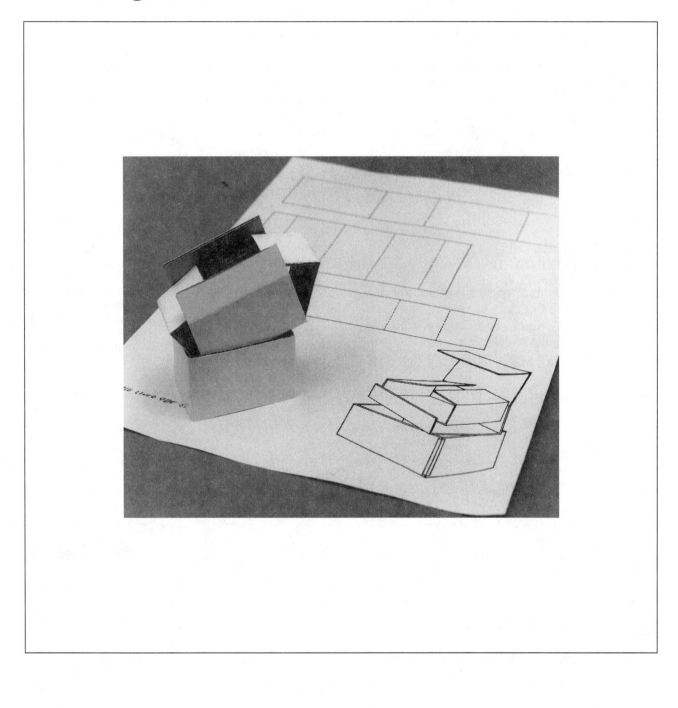

Fluted medium, the heart of the corrugated board, is one of the most unusual of all packaging materials. Originally it was an article of wear. In the mid-nineteenth century, men's hats were fashioned with a sweatband of fluted paper. In 1871 an American, Albert L. Johnes, patented fluted medium for the protection of bottles during storage and shipment. Three years later another American, Oliver Long, patented a process for sandwiching the fluted medium between paperboard sheets. This was the origin of corrugated containers, the workhorse of the packaging industry.

THE STRUCTURE OF CORRUGATED BOARD

The basic structure of corrugated board is simple. It consists of a fluted sheet glued to one or more liners. The most common construction is a sheet of "corrugated medium" sandwiched between two liners. A wide variety of combinations are possible, depending on packaging requirements. Where great strength is required, the three sheets of medium can be combined with appropriate liners.

The structural characteristics of corrugated board are governed by four variables: (1) the strength of the liners, (2) the strength of the corrugated medium, (3) the height and number of flutes per foot, and (4) the number of walls (single, double, or triple).

Simplified schematic of a corrugator. The corrugating medium is preconditioned and shaped by the corrugating rolls. The preheated linerboard is then glued to the face of the corrugated paper and conveyed to the double facer for attachment of the liner on the opposite side. The assembly is then dried over hot plates and run through pressure rolls for stability. Once the medium has passed through the cooling section, it is cut to the required size at the delivery section.

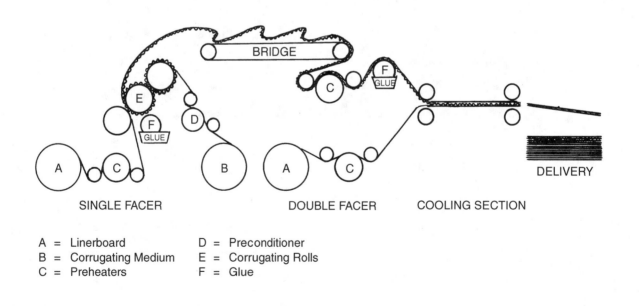

A = Linerboard   D = Preconditioner
B = Corrugating Medium   E = Corrugating Rolls
C = Preheaters   F = Glue

A corrugated flute gauge (actual size).

Unlined corrugated (A Flute)

Single Face Corrugated (A Flute)

Single Wall (Double Face) Corrugated (A Flute)

Double Wall Corrugated (A and C Flute)

A number of flute structures are available, depending on packaging specifications. *A-flute* has great capacity to absorb shocks owing to the wider spacing of the flutes. *B-flute*, because of the larger number of flutes per foot, provides maximum crush resistance. *C-flute* combines the properties of the first two types, and *E-flute* is used where very thin corrugated board is recommended. In addition, a new grade of corrugated board has recently been developed which has a series of uniquely formed grooves and ridges that are smaller than those in the E-flute and which has 15–20% more ridges per lineal foot than the E-flute.

So universal is the use of corrugated board that its production is viewed as a barometer of the economy as a whole. It is hard to imagine products that cannot be packaged and shipped in corrugated boxes. Today more than 1,160 products—including live fish!—are shipped in corrugated boxes.

Corrugated board is used in multicolored shippers and point-of-purchase displays. It is among the least expensive of all packaging materials.

Corrugated Containers

A Flute Corrugated—33 flutes per linear foot. Approximately ³⁄₁₆″ (without thickness of facings).

B Flute Corrugated—47 flutes per linear foot. Approximately ³⁄₃₂″ (without thickness of facings).

C Flute Corrugated—39 flutes per linear foot. Approximately ⁹⁄₆₄″ (without thickness of facings).

E Flute Corrugated—90 flutes per linear foot. Approximately ³⁄₆₄″ (without thickness of facings).

DESIGNING A CORRUGATED BOX

The design of a corrugated box is a major undertaking. The process of selecting the correct package design for a particular product has grown more complex as new technologies and materials present ever-increasing manufacturing options.

The ways in which corrugated board may be used are practically limitless. Certain basic container styles and designs are suitable for packaging a wide range of products. There are some corrugated interior devices (platforms, padding, or inserts) and plastics (molded polystyrene foam) used to provide reinforcement, bracing, and shock absorption. These are illustrated and described on pages 469–506.

Specialty containers are tailored to the requirements of a particular product. Those requirements may involve everything from the "shipability" of the product itself to how the container is filled, stored, loaded, stacked, braced, dropped, and unpacked. Specialty or custom-made boxes are usually required for special products in large quantities (10,000 or more). Master cartons, which are shipping cartons that hold smaller cartons, are used for food, detergents, housewares, and hardwares.

Government and industry standards and regulations are designed to protect the users of cartons. There are laws pertaining to method of shipment, such as rail, air freight, truck, and parcel post (U.S. Postal Service). In addition, all corrugated materials and cartons must be certified by the manufacturer. Weight, paper content, and puncture and bursting test certificates must be displayed on all corrugated boxes.

A significant trend in corrugated technology is impregnating and coating corrugated board with waxes and plastics. The moisture-resistant coating permits reuse of the carton and shipment of products, such as produce, that were previously shipped in expensive wooded crates and barrels.

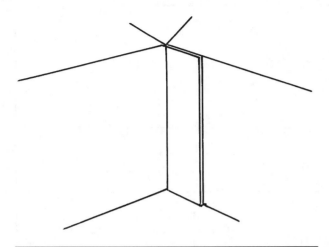

Glued joint: the manufacturer's joint tab is glued to the inside of the box.

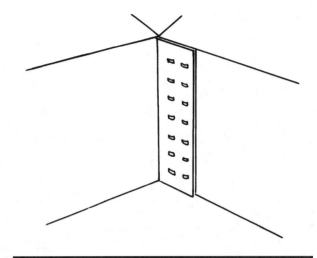

Stitched joint: the lap is stapled to the outside of the box.

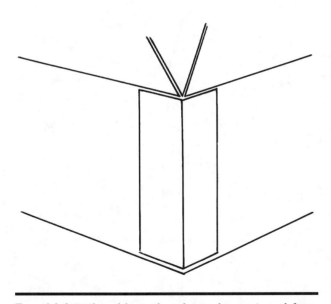

Taped joint: the side and end panels are taped from the outside of the box. No manufacturer's joint tab is needed.

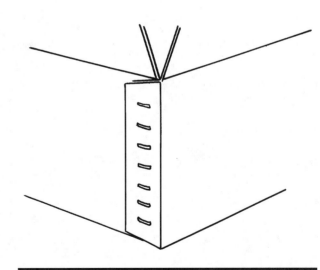

Stitched joint: the lap is stapled to the inside of the box.

Corrugated Containers

Corrugated Containers

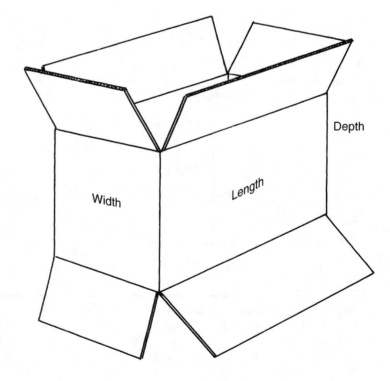

PRINTING ON CORRUGATED BOARD

Direct printing on brown corrugated board, which has a highly absorbent surface, is usually limited to the use of line art. This is the least expensive type of printing usable on corrugated board. Letterpress and, more recently, flexography are the typical printing methods.

Preprint is a term used to refer to the process in which a roll of printed stock is used as the top liner in making corrugated sheet. Flexography and rotogravure are the printing methods used in most preprint processes. The surface may be kraft, white clay-coated kraft, clay-coated bleached liner, or foil. The set-up costs for preprinting are high.

It is difficult to print full color on most corrugated board, with the exception of E-flute. Therefore, labels are often prepared to cover the boxes on one or more of their sides: full-sized labels cover the top and all four sides of a box; partial labels may be used only on the top of a box or on one or two of its sides. There are several variations of label application, including lamination onto the box, depending on size, shape, and cost considerations.

Litho labeling refers to lithographic printing of a sheet of paper that is then adhered (laminated) to corrugated board. Labels can be full or partial. Litho labeling and laminations are used on large boxes for toys, games, housewares, and sporting goods.

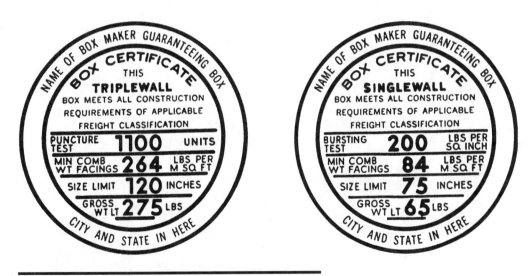

Weight, paper contents, and puncture and bursting test certificates must be displayed on all corrugated containers.

OVERLAP SLOTTED CONTAINER
(Intl. Box Code 0202)

417

Corrugated Containers

Corrugated Containers

Corrugated Containers

420

Corrugated Containers

Corrugated Containers

CENTER SPECIAL OVERLAP SLOTTED CONTAINER
(Intl. Box Code 0205)

Corrugated Containers

Corrugated Containers

Corrugated Containers

Corrugated Containers

Corrugated Containers

DOUBLE COVER BOX
(Intl. Box Code 0310)

#2 BLISS BOX
(Intl. Box Code 0605)

435

Corrugated Containers

Corrugated Containers

This eight-count two liter soft drink carrier uses a minimum of corrugated board and serves as a display unit.

Grain Direction

Doubled-over inner walls, hinged from the top, secure side panels in place. The two layers of the bottom and front wall as well as the triple sidewalls with carrying handles produce an excellent container for files.

DOUBLE END WALL SELF-LOCKING TRAY
(Intl. Box Code 0422)

This sample has supports for a U-shaped lightbulb.

Corrugated Containers

This tray or caddy with both wings and ears, and Walker Locks features a platform off the tray's front panel with an audio or video cassette well.

A pad cut from the waste of the sheet can provide a floor for the cassette well.

This protective frame corner is adjustable to three different frame thicknesses.

Corrugated Containers

Corrugated Containers

Corrugated Containers

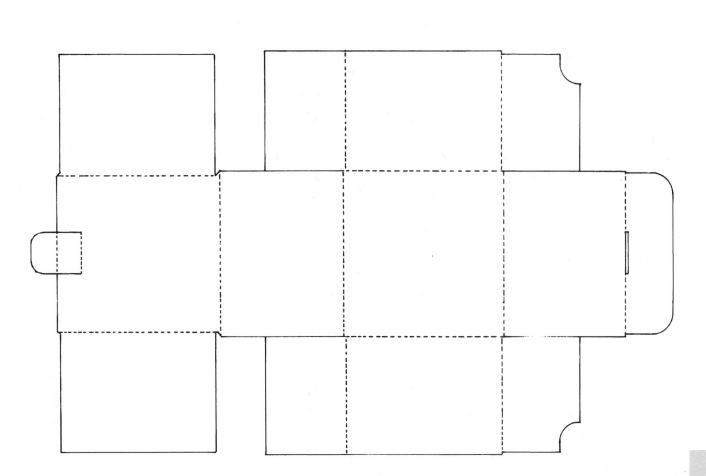

This blank folds into a sturdy container for heavy products, in this case a 5 lb. spool of picture wire.

Corrugated Containers

Developed by contemporary designer John Edminster, this structure provides a decorative, reclosable three-faceted top for a hexagonal carton.

Corrugated Containers

467

This carrying case requires no glue, features interesting configurations for the graphics, and is suitable for a large number of products.

Grain Direction

Corrugated Containers

Corrugated Containers

Corrugated Containers

Corrugated Containers

Corrugated Containers

Corrugated Containers

Corrugated Containers

Corrugated Containers

Corrugated Containers

Corrugated Containers

Corrugated Containers

Grain Direction

Corrugated Containers

This protective cushioning device may be die-cut to conform to many different fragile objects.

Corrugated Containers

Corrugated Containers

Corrugated Containers

Corrugated Containers

Corrugated Containers

Corrugated Containers

Corrugated Containers

5

Point-of-Purchase Displays

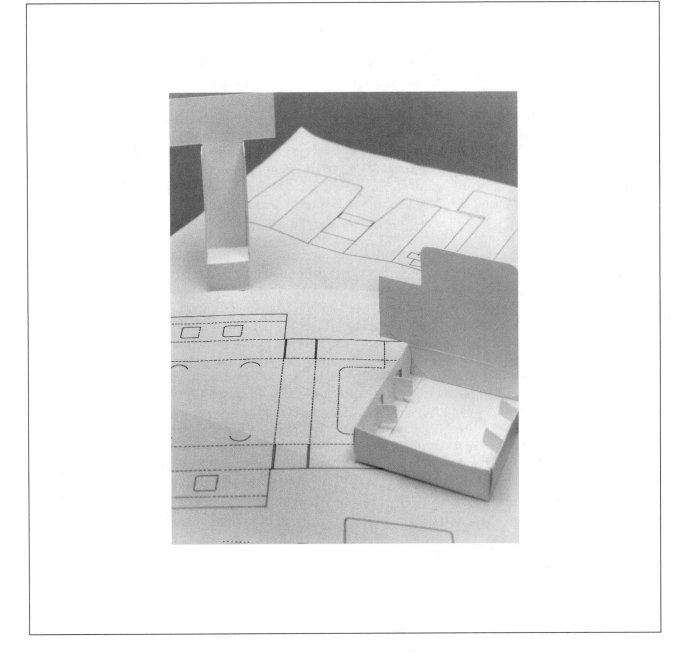

Die-cut three-dimensional displays first appeared in 1910, and photographic full-color displays made their appearance in the late 1920s. With the development of new plastic materials and techniques, point-of-purchase (P.O.P.) displays can now be produced in a variety of styles for practically all consumer products. With the rapid growth of many industries and the increasing variety of products, the point-of-purchase display has become an ever more effective selling aid for retailers.

The expansion of self-service stores and changes in consumer buying habits have also contributed to the development of P.O.P. materials. The modern retail establishment is a busy place in which a large percentage of selling is accomplished through self-service; consumers choose merchandise themselves rather than having it brought to them by salespeople. Often, unplanned buying decisions are made in the store, and an effective, well-designed display may be the deciding factor in a consumer's choice.

Planning and designing a P.O.P. display is a complex job involving a variety of materials and technologies. Paperboards, plastics, and corrugated boards all play an important part in the production of the display.

There are several categories of P.O.P. displays. Among them are display merchandisers, promotional displays, display shippers, counter displays, floor displays, and gravity-fed displays.

(Large supermarket displays, permanent displays, wire displays, dump bins, and motion displays, all of which tend to be larger and more expensive to produce, are not discussed here.)

Display merchandisers play an important part in self-service systems because they are strategically placed in the store, usually near the cash register or checkout counter. The display merchandiser is sometimes called a promotional display because it is designed to be used only for the duration of a particular promotion.

Promotional displays have a short life, usually three to four weeks (i.e., the duration of the promotion). The material used is usually paperboard or E-flute corrugated, often combined with an inexpensive vacuum-formed (thermoformed) plastic platform to hold the product.

The *display shipper* is the most common variation of the promotional display. It is a shipping carton that opens up to form a display with a die-cut *riser* (or reader) panel for art and copy. Shippers are used in the mass merchandising of health and beauty aids, liquor, toys, novelties, pharmaceuticals, and books. The advantage of the shipper is that it combines a shipping carton with a display setup.

Gravity-fed displays are one of the oldest types of display systems. They are used for film, shoe polish, and small packages of various products.

The following pages contain a variety of structural designs and patterns for P.O.P. displays.

Point-of-Purchase Displays

This S.W.E. (single wing easel) invented by Mike Scotti
stays erect by spring tension.

The ears slide in the slots when the easel is set up.

May be locked by pop-ins on the bottom panel.

May be glued to back of cards, boxes, etc., or may be keyed with lock tabs to display.

517

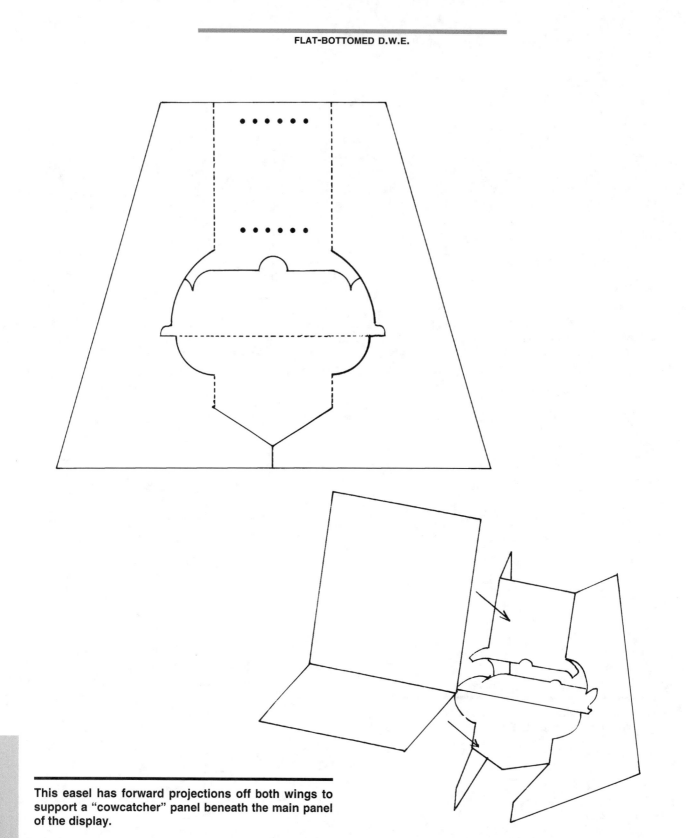

This easel has forward projections off both wings to support a "cowcatcher" panel beneath the main panel of the display.

Point-of-Purchase Displays

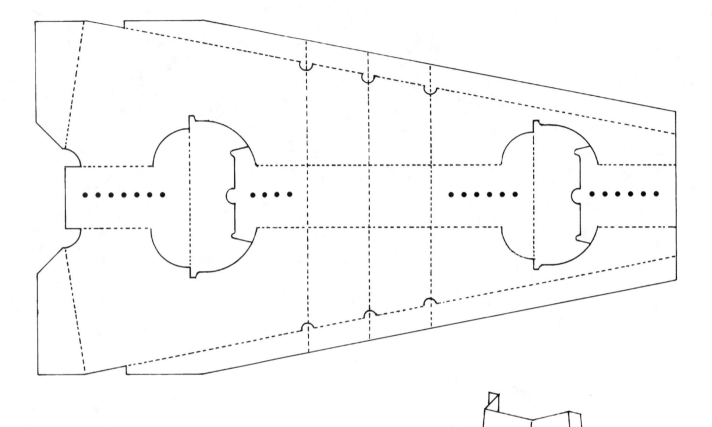

This easel is especially suitable for very tall displays that need to be folded in half for shipping. The middle portion of the easel may be accordion-folded back out of the way of the main lateral crease of the display.

The reinforcing turnbacks lend extra stability.

ONE-PIECE, DOUBLE SIDE WALL, TILTED-PLATFORM DISPLAY
STAND WITH BUILT-IN EASEL

523

This is a rigid (Set-Up) box; the corners of the outer tray are connected with stay tape and the tray is covered with cover stock.

Point-of-Purchase Displays

The stacking tabs that protrude above the top surface of the container insert into the slot of the above-stacked container with a compatible footprint.

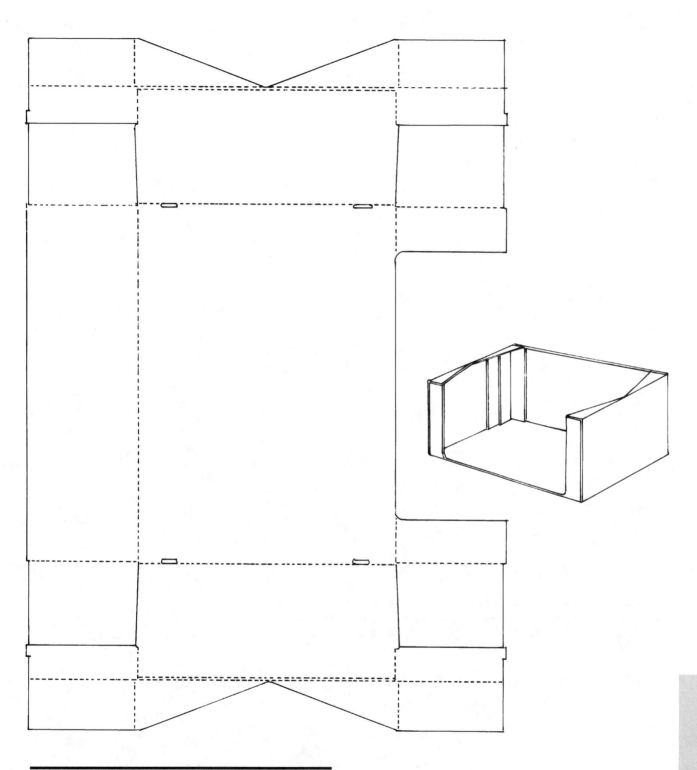

Shown with top and partial front removed, this unit has
reinforced corners.

Point-of-Purchase Displays

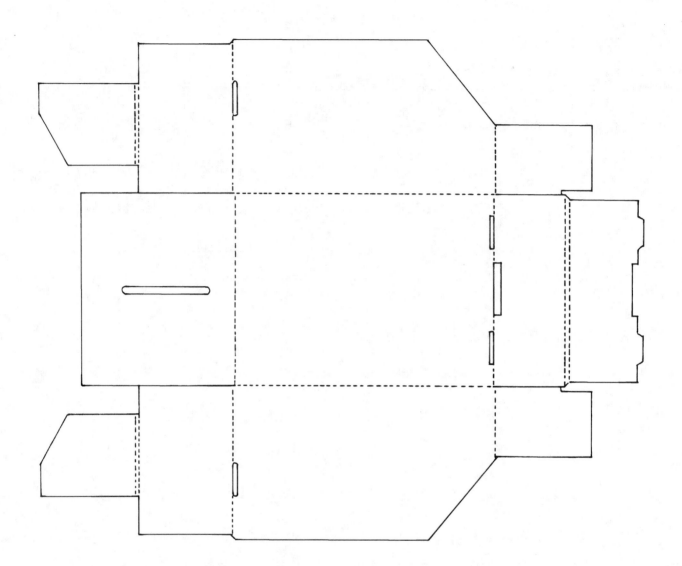

As a shipping container this carton features a lid attached to the back panel (not shown). Once the lid has been torn off, the unit acts as a display bin.

Point-of-Purchase Displays

The lid is prescored and perforated to create a display panel when opened.

The perforated panel doubles as a header.

Assembly requires no glue.

This is a patented type of display construction.

This is a patented variation of the rhomboid type of display container.

The slits in the inner walls permit placement of carded products.

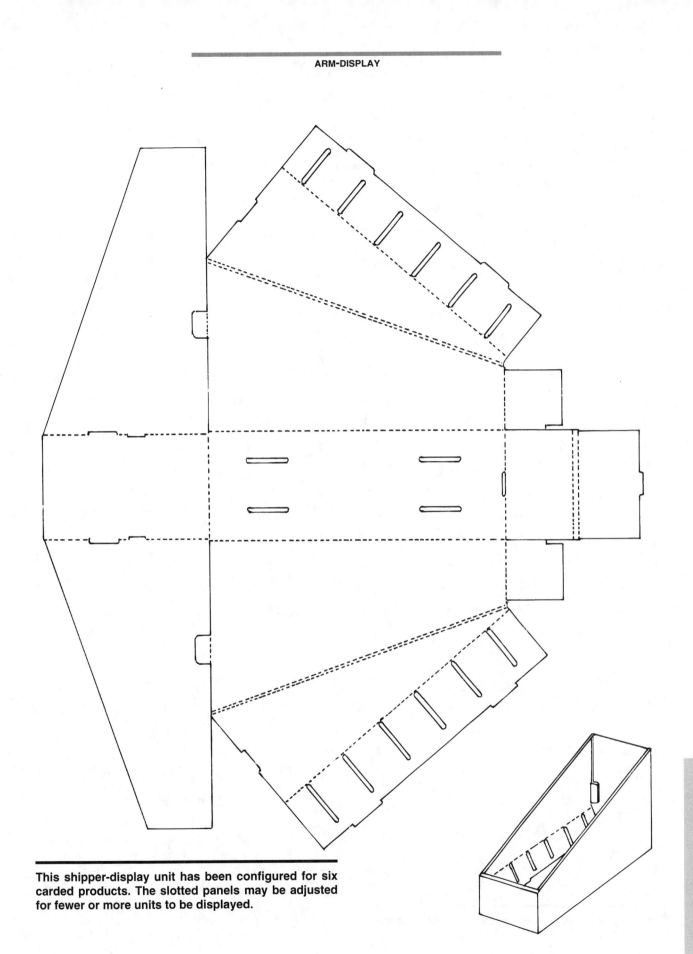

This shipper-display unit has been configured for six carded products. The slotted panels may be adjusted for fewer or more units to be displayed.

This two piece container has an insert which is configured for carded merchandise.

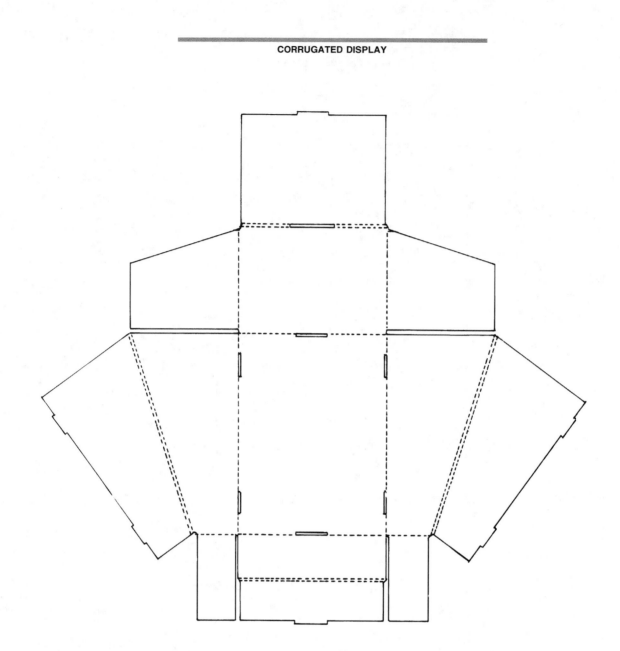

This double front wall, angled double side-wall tray with Walker Locks has a slit in the top of the double back wall to accommodate the separate header or riser illustrated on the next page.

The header may be die-cut into any suitable configuration; the "tongue" has been formatted to fit the slit in the back panel of the corrugated display.

The insert helps the contents of the display lean back.

Point-of-Purchase Displays

Ideal for a "dump" style of display. A base may be
added to make this a floor display.

Point-of-Purchase Displays

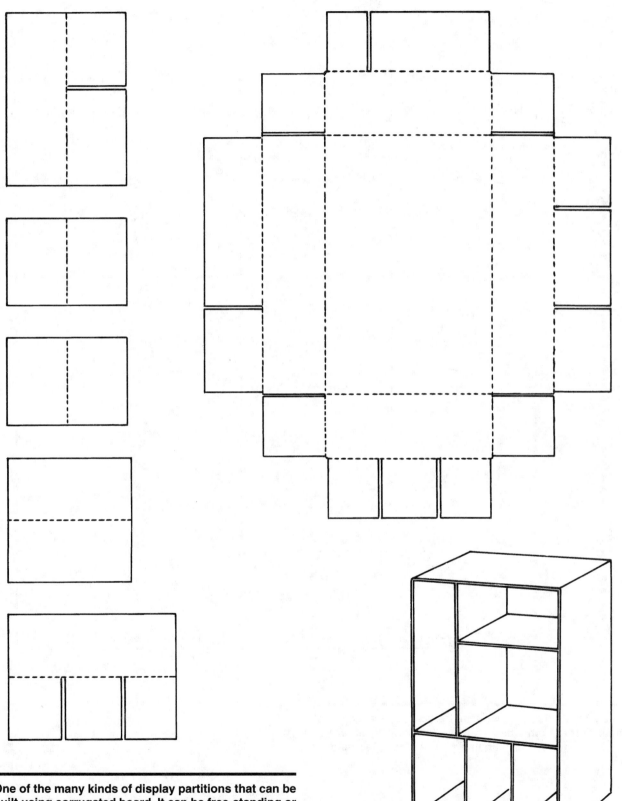

One of the many kinds of display partitions that can be built using corrugated board. It can be free-standing or supported by the display stand on previous page.

Point-of-Purchase Displays

Flute Direction

This 50" tall display base may be scored vertically in the center of both side panels and horizontally across all panels for easier shipping and storing. (See page 558.)

Three shelves

Two panels

Two of these units fit on the display base pattern on the preceding page. The protruding ears on top of the unit fit into the base of the top bin.

Point-of-Purchase Displays

Corrugated trapezoidal display base. The lateral score lines in the side and back panels indicates the *Breakdown Score* which allows the base to be folded in half for shipping.

Flute Direction

The slot in the bottom front panel mates with the tongue on the inside-out shipper on the following page.

Point-of-Purchase Displays

The shipping container has the graphics on the inside panels. The bottom is taped; top panels are folded back and the riser panel is kept in place by the tongue and slot combination.

Flute Direction

The ears on the front panel lock into the slots on the side panels. The slots in the front panel and ears on the side panels match up with the inside-out shipper/display container on the following page.

Point-of-Purchase Displays

The printed inside of this container unfolds into a display tray at point of sale. The riser panel's tabs fit between the side panels and the cover flaps that are folded back. The slots and ears mate with the ears and slots of the base on the previous page.

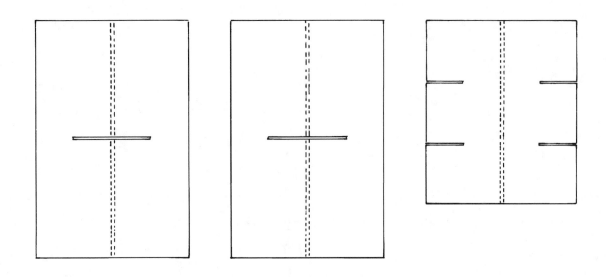

Display containers can be divided into sections fitted to the product(s) to be displayed. The corrugated dividers shown here are slotted, folded, and slide together to form six equal partitions.

Point-of-Purchase Displays

This corrugated tray with double front and side walls can be used as a counter display or as one level of the "waterfall" display shown on the following page.

Flute Direction

The 3-step waterfall corrugated display stand is ideal for showing one or more products on different levels.

Flute Direction

This pattern is but one of the many configuration possibilities. The top closure panel may be folded upon itself to add header space.

The slotted and scored insert carries the weight.

May be used by itself or as a base.

The scored and folded back base supports the platform.

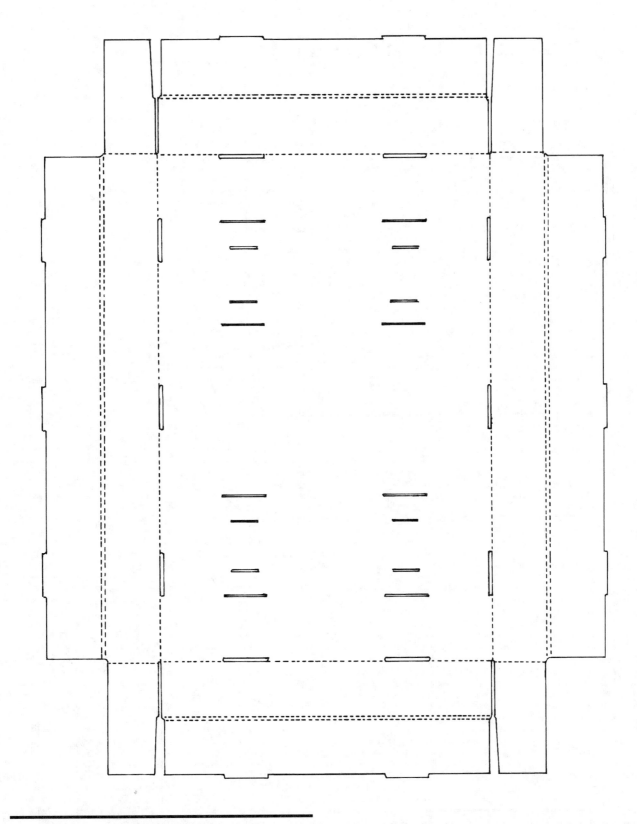

The corrugated speed table top features slots to accommodate the belt locks of the two stands.

A pair of these units is required for the speed table. The belt locks secure the top to the stands and the slots mate with the support section.

Belt Lock

The slots in the support mate with the two leg sections. The top is then secured to the stand with the eight belt locks.

This version has inner well walls that click into the slots in the bottom. One of the four walls imparts a slight lean-back for the separate riser.

Point-of-Purchase Displays

Altering the inner well walls permits the sleeve to fit snugly in this display. This pattern lends itself as well to a paperboard container; a second unit could serve as a closure when placed upside down over the sleeve.

Glossary

Adhesion   The process of attaching objects (e.g., by pasting, gluing) with adhesive materials.

Aseptic packaging   Sterilized containers made of plastic-lined paper (foil and plastic lamination).

Base   The lower or receptacle portion of a set-up box.

Blank   A folding carton after die-cutting and scoring but before folding and gluing.

Blind embossing   See *embossing*.

Box   A complete paper box, including base and lid.

Brightwood box   A tray-style carton made on the Brightwood machine.

Bundle   A unit of boxboard containing 50 pounds of sheets.

Bursting strength test   A test that measures the degree of resistance of a material to bursting.

Calendering   A process in which paper is pressed between metal rollers to form a continuous sheet under controlled conditions of speed, heat, and pressure.

Caliper points   A measure of paper thickness, expressed in units of a thousandth of an inch and usually written in decimals.

Carton   A unit container made for bending grades of boxboard.

Chipboard   Recycled paperboard.

Compression test   A test in which force is applied by two flat surfaces to opposite faces of a box.

Converter   A manufacturer that fabricates folding cartons from boxboard and other packaging materials (e.g., sheet plastics).

Corrugated line board and medium test   A test that measures the amount of force required to crush a fluted corrugated medium; measures the stiffness of the material.

Corrugated paperboard   A material constructed from alternating layers of fluted paperboard. (Flutes are classified as A, B, C, and E.)

Count   The actual number of sheets of a given size, weight, and caliper required to make a bundle weighing 50 lbs.

Cutting and creasing die   A steel rule form used on a press to cut and score sheets of boxboard into folding carton blanks.

Die-cutting   The process of cutting shapes from paperboard using cutting or stamping dies or lasers.

Dividers   Loose strips of paper or boxboard used to separate items of merchandise.

Double wall   A material consisting of three flat facings and two intermediate corrugated members.

Drop test   A mechanical procedure used to test the safety of package contents during shipping; determines the resistance of a filled box to shocks caused by dropping it in certain ways (i.e., on corners, edges, faces, etc.).

Easels   Supports attached to a box to hold it upright for display.

Embossing   A process in which paper is pressed between metal dies to create an image in relief.

Fiberboard   A general term applied to fabricated paperboard utilized in container manufacture; may be either corrugated or solid.

Filled board   Boxboard containing center plies consisting of different materials from those used for the top and bottom liners.

Flap lid   A lid that has no sides or ends and is hinged to the base of the box.

Flexographic printing   Printing from flexible plates made of rubber or plastic.

Flutes   The wave shapes in the inner portion of corrugated fiberboard. The flute or corrugation categories (A, B, C, and E) indicate flutes per linear foot.

Folding carton   See *carton*.

French lid   A lid with an extension edge, sides, and ends; it fits outside the base of the box.

Furnish   A mixture of water, pulp, paper scrap, sizing, dyes, and other additives that is fed into the wet end of a paper- or board-making machine.

Grain   The direction in which the fibers in paper line up.

Gravure printing   A printing process in which specially etched cylinders with cells to accept and store inks rotate the image to the impression cylinder, which transfers the image to paperboard.

Inner packing   Materials or parts used in supporting, positioning, or cushioning an item.

Labels   Die-cut, often self-adhesive applications used to decorate or identify packages.

Laminating   Bonding together two or more layers of material, usually impregnated with thermosetting resin.

Laminator   A machine that adheres two or more plies of paper or board.

Laser   A device that produces a very narrow beam of extremely intense light; used in industrial processes, medicine, and detailed die-cutting.

Letterpress   A printing method in which ink is transferred from a metal plate directly to a sheet of paperboard.

Lid   The upper and covering portion of a paper box.

Lifts   Strips of ribbon, cloth, tape, or paper attached to an inner tray, allowing it to be easily removed from the base.

Loose wrapping   A process in which a lid is covered with paper on a wrapping machine or by hand; the wrapper is a single piece with adhesives only at the edges.

Manufacturer's joint   The seam of a carton, where the two edges of the box are joined together by stitching, gluing, or taping.

Master carton   A carton that is used to ship smaller cartons.

Neck   A shell inserted in the base of a shoulder box, attached by an adhesive and extending above the base into the lid.

Newsback   Chipboard that is lined on one side; used for inexpensive printing.

Offset printing (lithography)   A printing method in which ink is transferred from a plate to a smooth rubber "blanket" roller that transfers the image to paperboard.

Padded top   An extra top, covered with paper or cloth and including one or more layers of wadding or other padding material, that is attached to a top lid.

Paperboard (cardboard)   A material made from laminated layers of paper in sheets of 12 points or more.

Paper stock   Reclaimed (recycled) material that has been pulped for the use as a finish for paperboard; includes newsprint, reclaimed corrugated containers, and mixed paper and boxboard (kraft) cuttings.

Paster   A machine that applies an adhesive to two or more plies of paperboard.

Ply   A layer of boxboard formed on a multicylinder machine.

Pulp   Basic cellulose fibers resulting from the disintegration of wood, rags, and other organic materials by chemical or mechanical processes; used to make boxboard.

Puncture test   A test of the strength of material, expressed in ounces per inch of tear, as measured by the Beach puncture tester.

Recycling   The use of previously used materials to make new objects.

Scoring   Making an impression or crease in a box blank to facilitate bending, folding, or tearing; feeding boxboard sheets through a scoring unit.

Set-up paperbox   A paperbox or rigid construction that has been formed or set up and is ready for use (as opposed to a folding carton or shipping container; see corrugated box).

Shell   A plain, unprinted carton designed to be overwrapped with plain or printed paper, fabric, or other materials.

Shoulder box   A box with a glued neck inserted so that its base, ends, and sides form a shoulder on which the lid rests.

Slide box   A box in which the lid is the form of a shell into which the base is inserted at the side or end.

Stay   Material used to reinforce the corners of a base, lid, or tray; may be paper, cloth, combination (cloth and paper), wire, metal, or plastic tape.

Stitching, stapling, and taping   The application of metal fasteners or tapes to form or close a box.

Telescope box   A box in which the sides and ends of the lip slip over the base to either partial or full depth.

Tight wrapping   The process of covering a base, lid, or tray with paper on a wrapping machine or by hand; the wrap is a single piece whose entire surface is covered with adhesive.

Tray-style carton   A folding carton with a hinged cover extending from the side walls.

Trim   Paper or cloth covering to strengthen or decorate base edges and lid or extension edges.

Triple wall   A material consisting of four flat facings and three intermediate corrugated members.

Tube-style carton   A folding carton that encloses a product with flap, tucks, and locks.

Tuck   The end portion of the top of bottom flap of a folding carton, which is inserted into the container to hold the end flaps in place.

Universal Product Code (UPC)   A printed code on containers and other forms of packaging that provides information about the product for purposes of inventory control and retail pricing.

Vibration test   A method of checking vibration during transportation.

Water absorption test   A measure of resistance to water absorption; applied to boxes impregnated with water- and moisture-resistant coatings.

Web   A roll of paper or other flexible material as it moves through the machine in the process of being converted.

Bibliography

Abbott, Donald L. *Packaging Perspectives*. Dubuque, IA: Kendall/Hunt, 1989.

Brody, Aaron L. and Kenneth S. Marsh (eds.). *The Wiley Encyclopedia of Packaging Technology, 2nd edition*. New York: John Wiley & Sons, Inc., 1997.

Fibre Box Handbook. Chicago, IL: Fibre Box Association, 1986.

Griffin, Roger C., Jr., and Stanley Sacarow. *Principles of Package Development*. Westport, CT: AVI, 1972.

Handbook of Folding Carton Style Nomenclature. Washington, D.C.: Paperboard Packaging Council, 1988.

Hanlon, Joseph F., R. J. Kelsey, and H. E. Forcino. *Handbook of Package Engineering*. Lancaster, PA: Technomic Publishing Company, Inc., 1998.

Hine, Thomas. *The Total Package: The Secret History and Hidden Meanings of Boxes, Bottles, Cans, and Other Persuasive Containers*. Boston: Little, Brown & Company, 1997.

Packaging Encyclopedia and Technical Directory 1989. Des Plaines, IL: Cahners, 1989.

Index